CANYONLANDS

COUNTRY

Text, Illustrations, and photographs by Don Baars,
unless otherwise indicated.

Cañon Publishers Ltd.

411 LAWRENCE AV.
LAWRENCE, KS
66049   USA
TELEPHONE 913-842-0570

AND

Canyonlands Natural History Association
125 West 200 South
Moab, Utah 84532
(801) 259-6003

FRONTISPIECE: South-facing escarpment of Island in the Sky, from near the foot of
Shafer Trail.

GEOLOGY OF CANYONLANDS COUNTRY

TABLE OF CONTENTS

PREFACE

There is a need for a book of this kind; one that explains the innermost geological secrets of Canyonlands and Arches national parks in southeastern Utah, in language understandable to non-geologists. At this writing, nothing of the sort is available to the curious visitor.

Geology is a science, of sorts, but in actual practice is more like a detective story. We really have few clues to work with, yet we try to make a reasonably sensible story out of what we have. Consequently, there are sometimes as many interpretations as there are "detectives" working on the case. That is true in Canyonlands Country, but this story is as factual and as correct as one opinionated geologist can make it.

For convenience, this book is divided into two parts. Part One is a general discussion of the Colorado Plateau region, an introduction to some of the applicable principles of geology, and a summary of the geologic history of Canyonlands Country. It is designed to give the interested reader sufficient geological background to appreciate the idiosyncracies of this unique landscape. We will first establish the geographic boundaries of Canyonlands Country and the Colorado Plateau, a physiographic and geologic "Province" in which our area of interest lies. Then we will explore the long and complex saga of the construction of the rock lattice of Canyonlands Country from bottom to top, from the first events until the latest. Finally, we will discover the tireless processes of erosional sculpting of the landscape, tearing asunder the rock layers so carefully emplaced. This part is perhaps best read before visiting the region's parks.

Part Two is a geological tour guide that may be used by itself or with the geologic history presented in Part One. Although it is not prerequisite, the generalities presented in Part One help explain the more detailed discussions of specific landscapes. This section deals with local features seen at the surface and does not delve into background material. It is designed to be used while touring the various districts of the parks, and is best understood after reading Part One.

To fully appreciate the geology of this magnificent region, it is strongly recommended that the "Geologic Map of Canyonlands National Park and Vicinity, Utah" and the "Geologic Map of Arches National Park and Vicinity, Grand County, Utah" be used in conjunction with this book. The waterproof maps published by Trails Illustrated are helpful in navigating the backroads and trails of Canyonlands Country, and the waterproof river guide "Cataract Canyon and Approaches" published by Cañon Publishers, Ltd. is recommended for river runners. For the novice to geology, "Scenes of the Plateaulands" by William Lee Stokes is highly recommended reading. These publications are available at the two parks, or from Canyonlands Natural

History Association in Moab, Utah.

Many people have aided materially and spiritually in the development of this book. Canyonlands Natural History Association provided endless encouragement and financial support through a pre-publication purchase arrangement. Eleanor Inskip, Executive Director of the Association, acted as intermediary, keeping the project on an even keel; her efforts and continuing cheerfullness were foremost in bringing this complex effort to completion. Dave May, Dee Tranter, and Pete Parry, Trustees of the Association, provided many helpful suggestions, keeping the contents of the book as truthful and readable as possible. Bill McDougald and Bob Norman, professional colleagues from Moab, added many helpful historical and geological suggestions. The editorial expertise of Rose Houk and Michaelene Pendleton improved the manuscript significantly, adding vastly to the clarity and readability of the text. Helpful suggestions by Jean Eardley, Lynn Jackson, and Saxon Sharpe are gratefully acknowledged. And finally, the loving encouragement and patience of my wife, Jane, made the book possible.

Jerry Rumburg became the Chief Interpreter for Canyonlands National Park as this book project was in its infancy, and he gave his enthusiastic support to the project. His untimely death in an automobile accident on December 16, 1987 deeply shocked those of us who enjoyed and admired his professional enthusiasm. His preliminary efforts to provide educational displays to Park visitors was appreciated and applauded by us all. This book is dedicated to the memory of Jerry Rumburg and his heartfelt goals.

★　　★　　★　　★

PART ONE
GEOLOGIC HISTORY

Double Arch

CHAPTER ONE
CANYONLANDS COUNTRY

"...The landscape everywhere, away from the river, is of rock - cliffs of rock; plateaus of rock; terraces of rock; crags of rock - ten thousand strangely carved forms. Rocks everywhere...

"When speaking of these rocks, we must not conceive of piles of boulders, or heaps of fragments, but a whole land of naked rock, with giant forms carved on it: cathedral-shaped buttes, towering hundreds or thousands of feet; cliffs that cannot be scaled, and cañon walls that shrink the river into insignificance, with vast, hollow domes, and tall pinnacles, and shafts set on the verge overhead, and all highly colored buff, gray, red, brown, and chocolate; never lichened; never moss-covered; but bare, and often polished." (John Wesley Powell, 1875)

This, one of the earliest written descriptions of Canyonlands, may still be the best. Such were the impressions of Major John Wesley Powell, recorded after his explorations of the Green and Colorado rivers in 1869 and 1871-72. Powell, a veteran of the Civil War who lost part of his right arm from an injury received at the battle of Shiloh, was a self-trained geologist with a penetrating curiosity and an indomitable spirit of adventure. When Powell and his men made their first river trip down the canyons of the Southwest, Canyonlands Country was only a large blank spot on the map. Even the geographic location of the confluence of the Green and Colorado (then known as the Grand River) was totally unknown.

Members of the Powell Expeditions were not the first white men to see the region. They had been preceded by others, the best known of whom was Denis Julien, a French-Canadian fur trapper who left his name and dates scratched on the canyon walls in 1836 and elsewhere in the 1840s. Julien's only claim to fame was his obsession for writing his graffiti on the cliffs; Powell's descriptions of the Canyonlands were eloquent, scientifically significant and equally as lasting.

The story behind the origin of the "ten thousand strangely carved forms" of Canyonlands Country is as strange and fascinating as the rugged landscape itself. A saga, developing for much longer than Major Powell ever could have imagined, has culminated in myriad crags, angular mesas,

Aerial view northward up the Green River from near the confluence with the Colorado River. The slopes just above river level are in the Elephant Canyon Formation. The lowest prominent bench of lower Cutler Formation is overlain by the Organ Rock Shale, capped by the White Rim Sandstone that supports the extensive White Rim bench. Cliffs in the distance are in the Moenkopi, Chinle, Wingate, and Kayenta formations in ascending order.

mushroom-shaped pedestals, and cliff-bounded plateaus separated from one another by sharp crevices, deep ravines, arroyos, and especially canyons; canyons of all sizes and shapes dominate the view. But, above all, as Powell said, of "...Rocks everywhere..."

To understand and fully appreciate this magnificent landscape, we must look back into time some 25 or 30 million years, when the Green and Colorado rivers began to carve the canyons we admire today, and to a more distant time, some 60 or 70 million years ago, when the last episode of rock-folding and mountain-building took place. We must understand events of 300 million years ago when thousands of feet of rock salt that now lie beneath this land were deposited in a vast "dead sea." And to understand these relatively recent events, we must examine the development - nearly two billion years ago - of great, deep-seated fractures in the Earth's crust.

The lengthy and involved history that we now know to be the story behind the scenery in Canyonlands Country was beyond the grasp of early pioneer geologists. Though their explanations appear today to be oversimplified, the story that has evolved from more than a century of inquisitive study began with Major Powell and the various government surveys of the late 1800s. Development of the story resulted from tireless hours and even years of searching for answers by hundreds of geologists who braved the rapids of the rivers, rode horseback, walked, and sometimes crawled or climbed over the inhospitable terrain, all in the search for mineral wealth or simply for the sake of knowledge. To list all these hardy souls would require a book in itself. Indeed, no one person is responsible for more than a minuscule fragment of knowledge in this continuing fact-finding mission.

Truly, today's geologists are standing on the shoulders of giants who have gone before, and have shown us the way.

KEEPING TRACK

The cliffs and buttes and canyons of Canyonlands Country consist of layers of sedimentary rock, each having its own topographic expression, color, and mood. Any particular layer is strikingly different from those above and below, yet it can be followed as far as the eye can see - and far beyond. Each layer of rock has a different geologic history and meaning; one is a stream deposit, another is an ancient field of windblown dunes, another was deposited in shallow, tropical seas. Because of this, it is important for the geologist to distinguish specific layers and to be able to communicate just which layer is being talked about. So we give each a unique name, much like people have individual names.

Most folks think of a silly-looking rock, perhaps shaped like an owl or an elephant, as a rock "formation." To a geologist, however, a "formation" is a layer, or series of similar layers, of rock that is geographically extensive, geologically significant, or both. Formation names are usually derived from a locality, or "type section," in which the layer(s) can best be studied. The name has two parts: the first part being a geographic place name, usually related to the type section, and the second part designates the kind of rock that typifies the formation.

For example, the rock layer that forms the massive, high cliffs surrounding Canyonlands and hosts the prominent view points, such as Dead Horse Point or Grand View Point, is named the Wingate Sandstone. The name "Wingate" comes from the type section of the formation at Fort Wingate, New Mexico, and the term "Sandstone" explains the rock type. The underlying slopes of varicolored shale are in the Chinle Formation, named for the type section near Chinle, Arizona; the term "Formation" means that several rock types are present. Even the name of that community is appropriate, for "Chinle" (pronounced chin-**lee**) is Navajo for "place where the water flows out of the mountain;" it is at the mouth of Canyon DeChelly (pronounced de-**shay**). No two formations have the same name.

The names of rock layers in Canyonlands Country are listed in the right hand column on the charts inside either cover of this book. They are listed in sequence from older at the bottom to younger at the top. Names used for distinguishing intervals of geologic time, terms used worldwide, are listed to the left. Major time intervals (Eras) are given in the left (first) column, smaller subdivisions of time (Periods) are in the second column. These are periods of time during which the named rock layers (right hand column) were deposited in Canyonlands Country. Estimated ages of the time periods in years before last Tuesday appear in the third column from the left. The origin and significance of these age names will be examined in Chapter Three. Notice that only the rock layers deposited later than the Middle Pennsylvanian are exposed here in Canyonlands Country. Older formations are known to be present only from deep drilling.

This method of deriving formation names hasn't always been so rigid. In the old days, a formation was sometimes named for a physical characteristic of the rock. For example, the Redwall Limestone in Grand Canyon was named because the gray limestone is coated with a "paint job" of red dirt washed down from overlying layers; thus the name "Red Wall" or "Redwall" Limestone. One couldn't get away with this in the present world of professional "regulations."

Then too, sometimes a formation is given a name, only to find out later that the rocks are too complex and need to be subdivided into more units. To accommodate this situation, rock units are ranked by their apparent importance, and the ranking can be changed to fit the needs of geologists. A "formation" can be elevated in rank to a "group" that can then be divided into two or more "formations." An example is the Hermosa Formation, named for rocks in Hermosa Mountain north of Durango, Colorado, but later found to change elsewhere to different kinds of rocks. The "formation" was upgraded to the Hermosa Group, and three separate formations were included. The upper two, the Paradox and Honaker Trail formations, are exposed in Cataract Canyon in Canyonlands National Park. They are of Pennsylvanian age, as shown on the chart inside the covers. Or a formation may be subdivided into "members" which may or may not be given individual formal names. The "Moss Back Member of the Chinle Formation" in Canyonlands is an example, but notice the complicated name that results. Member names are not included on the charts inside the covers because of space limitations, and are usually only distinguishable by a trained geologist with special interests.

All of this is designed to bring order out of chaos, and it usually works. The trouble is that a single layer of rock is often given separate names when studied in different areas. When this happens, the name first applied should be the one used. However, geologists are generally too provincial to follow the rule, and two names result for the same rock layer, especially if they outcrop in widely separated areas and their actual relationships were unknown at the time they were first studied. For example, the red Hermit Shale in Grand Canyon is the same bed called the Organ Rock Shale in Canyonlands. The situation was not understood when the formations were named and so there are two names for the same layer of rock.

The "art" of naming, correlating, studying, and keeping track of all this is called "stratigraphy" and those who worry about it for a living are called "stratigraphers." One wonders what these folks might look like.

8

CHAPTER TWO
THE COLORADO PLATEAU PROVINCE

The foreboding, desolate-seeming, but exquisite part of the world we call Canyonlands Country is south of the Book Cliffs and east of Capitol Reef in southeastern Utah. The region generally lies in the east-central part of the Colorado Plateau, a vast area geographers call a "province." A massive block of land, composed of less-deformed, relatively flat-lying and naked stratified rocks, the Colorado Plateau Province is distinctive for the general lack of geologic complexities typical of bordering regions. The provincial boundaries, marked by ancient, deep-seated structures that are expressed at the surface, are starkly clear in some areas but evasive and subject to personal interpretation in others. The boundary shown on the map on the opposite page approximates that used by geographers.

The western boundary of the province is marked by the Grand Wash Cliffs and High Plateaus in central Utah, and by the Wasatch Mountains near Salt Lake City. The northern boundary usually is considered to be the Uinta Mountains. The Mogollon Rim in central Arizona marks the western portion of the southern boundary of the province. From the eastern end of that obvious escarpment, however, the boundary steps eastward to Albuquerque, New Mexico, by any route you choose. Perhaps the best geologic boundary for the eastern limit of the Plateau country would be the valleys of the Rio Grande and Arkansas drainage systems, from Albuquerque northward to Alamosa and Leadville, Colorado. That, however, would surely upset those geographers who place the eastern limits west of the San Juan Mountains and Uncompahgre Plateau of western Colorado. Regardless, Canyonlands Country lies well within anyone's borders of the Colorado Plateau Province.

Although its name implies that the Colorado Plateau Province lies within the state of Colorado, most of the region is in Utah with only small parts occupying northern Arizona, northern New Mexico, and southwestern Colorado. The name is derived from the fact that the Plateau country is drained almost entirely by the Colorado River and its vast system of tributaries, the largest and most important of which are the Green, San Juan, Dirty Devil and Little Colorado rivers.

Elevations range from about 3,000 feet to over 14,000 feet, depending on how one draws the boundaries, averaging about 5,200 feet. Climatic conditions therefore range from Sonoran desert to alpine or arctic. Semiarid conditions generally prevail, enhancing the starkness of the rocky lands. Lower elevations are extremely hot and dry in summer; higher regions are exceedingly cold in winter. Moisture comes mostly in the form of brief, torrential rains during the summer months at lower elevations, where steady rains or winter snows are infrequent and short-lived. Higher elevations also are subject to summer thunderstorms, and winter snows may accumulate to

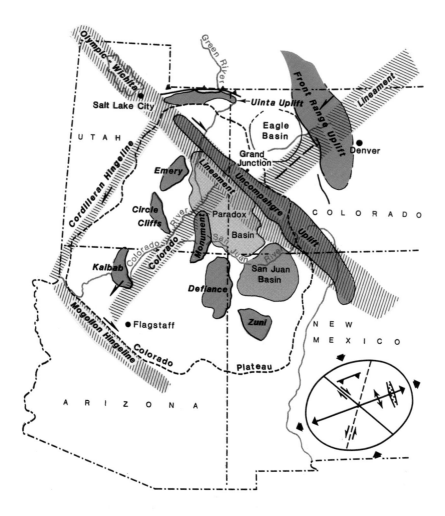

Map of the Colorado Plateau Province showing locations of the continental-scale lineaments (yellow) relative to the large uplifts (brown) and basins. The egg-shaped feature centered over southern New Mexico represents a block of rock, put in a vise, and squeezed from north to south (heavy arrows). Northwest- and northeast-trending wrench fault zones should develop with sense of movement as indicated by the smaller arrows. Compare these arrows with those along the major fault zones. East-west compressional folds (anticlines, such as the Uinta Mountains) should form (see long central double arrow in the "egg," and north-south extensional faults (such as those Precambrian normal faults that underlie the major monoclines of the Colorado Plateau) should form. Thus, this simple mechanism explains the "peculiar" orientation and location of the major structures of the Colorado Plateau Province.

depths of several feet. Periods of drought can be extensive and devastating to human and wildlife inhabitants.

Due to the lay of the land, highways on the Colorado Plateau have been built in the bleakest, drabbest parts of the region. This leads most visitors to believe that the country is a colorless wasteland to be avoided at all cost or, at least, traversed in the briefest possible time. The easiest places to build roads, although the hardest places to maintain them, lie on the Mancos Shale, a dark gray or black, soft-weathering rock which forms broad, flat valleys and, thus, easy routes.

Once one leaves the Mancos Shale wastelands, the country becomes a land of living rock, with unbelievably rugged topography and endless variation of magnificent colors. To the more adventurous who will brave the land beyond the reaches of the family station wagon, a wonderland of rugged, sometimes treacherous beauty and mystery unfolds. And, surprising as it may seem to some, this contrast in worlds is controlled by the geology.

REGIONAL GEOLOGY

The Colorado Plateau is a relatively high block of the continent that has acted independently of surrounding regions. It is set apart by large fault systems (long trends of Earth fractures, sometimes called "lineaments") that are major breaks in a continental-scale structural fabric that dates back to the Precambrian Era, more than half a billion years ago. Within this large block of thick Earth crust, the land has responded to numerous events of crustal unrest that have culminated in the rugged surface we see today. Hundreds of lesser deep-seated faults (fractures along which there has been some movement) have contributed to the formation of large uplifts and intervening downfolded basins in the rocks. Many of these structures are larger than some eastern states and some European countries, but are merely subdivisions of the Colorado Plateau Province.

THE PROCESS

If we think of the Colorado Plateau Province as a large chunk of solid rock placed into a gigantic vise, with the force of the vise's action oriented in a north-south direction, we would find that the rock would break along predictable cracks when we tighten the vise. Fractures would first develop as large X-shaped cracks in northeast-southwest and northwest-southeast directions, and cracks of lesser size would form in east-west and north-south directions. If we were to continue tightening the vise, movement would take place along the fractures and they would therefore become "faults."

Movement along the X-shaped faults would be more or less horizontal, creating "wrench faults" (strike-slip faults) that would offset the four triangular segments relative to each other. Meanwhile, vertical movement along the east-west faults would shorten the block from north to south (we can thus think of them as "compressional" features) and opposite movement

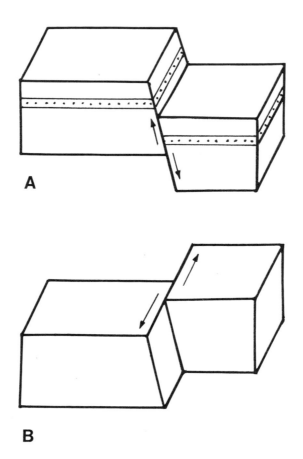

A

B

Block diagrams showing two kinds of faults commonly seen on the Colorado Plateau. Block A is a normal fault, where the right side of the block has slid down the fracture surface relative to the left side. Block B is a wrench fault (strike-slip fault), where the right side has moved laterally (horizontally) relative to the left side. Wrench faults usually occur in swarms, and are often very complex fault zones because of the tearing action involved where major blocks of rock move past one another. They are caused by severe compression or torsion of the Earth's crust.

along the north-south faults would lengthen the block from east to west ("extensional" features). And this is exactly how the Colorado Plateau Province was segmented nearly two billion years ago.

A thick coat of rubber cement placed over the original faulted block (to act as a cover of sedimentary rock) would buckle and wrinkle as we continue to tighten the vise. The old faults in the more brittle "basement" block would move in the same manner as before, and folds in the rubber cement would occur at the surface, concentrated over and along the original "basement faults." If we were to continue tightening the vise, occasionally adding another layer of rubber cement, the end result would be much like a miniature Colorado Plateau - with one exception. Late in our experiment we would have to take the broken chunk of coated rock from the vise, rotate it about 90 degrees, put it back in the vice, and put one more turn on the handle. Then we would have a true replica of the province.

THE EVENTS

To put this sequence of events into its proper time frame, about 1.7 billion (1,700,000,000) years ago, the province was strongly compressed in just this fashion. The fracture system described above was formed in the highly metamorphosed rocks (gneiss, schist, quartzite) and granitic rocks that now are buried deeply beneath the Colorado Plateau Province.

Time progressed, as it is wont to do, while accumulations of sediments blanketed the shattered basement surface to thicknesses of several thousand feet (our coats of rubber cement). Some sediments were deposited as layers of sand and mud in shallow seas at times of high sea level, some others as dune sands that covered the lands when the seas occasionally retreated, and still others as stream and beach deposits. Sand accumulations were cemented to form sandstones, muds were compressed to form shales, and the limy deposits formed in warm, shallow seas hardened into limestones.

Meanwhile, back at the vise, the screw was turning and north-south compressional forces continued their work, even as the sedimentary rocks were being deposited. The sedimentary rocks (our layers of rubber cement) were formed, gently folded, and occasionally faulted in the process. Thus, the structural fabric of the basement was propagated upward through the sedimentary sequence for a half billion years.

The notion that continent-size crustal plates scoot around and collide with one another over the Earth's surface is called "plate tectonics." It was on a sultry Tuesday afternoon in July, 70,000,032 B.C., that the worm turned, so to speak. More than 1.5 billion years of intermittent north-south compression ended as the North American plate changed course, sparks flying, and headed westward on a collision course with the oceanic plates of the Pacific.

This event, which geologists call the "Laramide Orogeny," had the effect of taking the block from the vise and rotating it. The "big squeeze" changed from north-south to west-to-east, with the result that old, gentle folds in the rocks were rapidly enhanced to major uplifts and basins, and structures were overturned sharply toward the east, forming the giant "monoclines" for which the Colorado Plateau Province is famous.

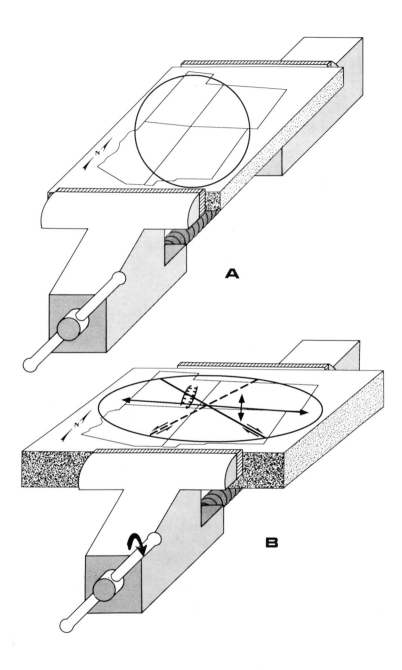

A

B

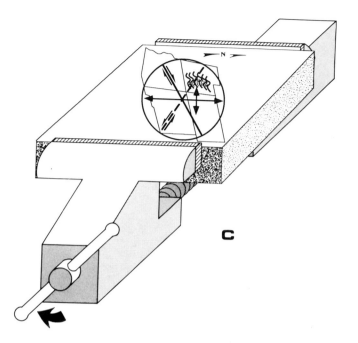

C

The Colorado Plateau, including large parts of the states of Utah, Colorado, Arizona, and New Mexico, has acted like a large, rigid block of the Earth's crust for nearly two billion years. This block, as shown in diagram A, was greatly compressed from north-to-south as if it were squeezed in a giant vise about 1.6 billion years ago. Compressive forces continued to shorten the Earth's crust for more than a billion years (through Permian time), causing major northwest- and northeast-trending faults to form as illustrated in diagram B. These wrench faults (faults with lateral rather than vertical movement) permitted shortening of the crust from north-to-south, and extension in an east-west direction (diagram B). North-to-south normal faults (faults with vertical movement that lengthen the Earth's crust) facilitated the extension as shown by the north-oriented barbed lines in diagram B.

In Late Cretaceous time, some 65 million years ago, Earth forces changed and compression came from the west, perhaps due to collision of the North American plate with the Pacific oceanic plate. It was as if the Colorado Plateau block were turned 90 degrees in the giant vise, causing shortening of the crust in the opposite direction (diagram C - note the 90 degree rotation of the north arrow and the state outlines). This formed the great monoclines of the Colorado Plateau, such as the Comb Ridge and San Rafael structures, where overlying sedimentary rocks draped over previously formed fault blocks. These folds are illustrated on diagram C by the wavy "wrinkle" lines. Thus, the ancient basement faults localized the much younger monoclinal folds we see at the surface today. [Illustrations by Brian Forgey]

15

MONOCLINES

Although this is intended to be a nontechnical description of Colorado Plateau geology, some strange terms simply cannot be avoided, the "-cline" family among them. Some "-clines" already are familiar to us: "inclines" are upward-sloping surfaces and "declines" are downward sloping. In geology, the important "-clines" are not necessarily surfaces, but involve considerable thicknesses of layered rocks. An "anticline" consists of rock layers that have been arched upward from their originally horizontal position, and a "syncline" is bowed downward. "Monoclines," the trademark of the Colorado Plateau Province, are a bit like half an anticline. The layers on either side of the monocline are more or less level, but they bend downward along the monocline like a carpet draped over a single step in a flight of stairs. How would a monocline be formed? Glad you asked! Simply have a deep-seated (basement) fault, buried by layers of sedimentary rocks, when compression in the Earth's crust forces one side (the up-thrown block) to move upward relative to the other side (the down-thrown block). Either or both sides may move, but in any case the overlying sedimentary layers will bend and drape over the edge of the up-thrown block onto the down-thrown block. Monoclines form some of the most notable features of the Colorado Plateau: for example Comb Ridge monocline forming the east flank of the Monument uplift, the Waterpocket Fold forming the east flank of the Circle Cliffs uplift, and the San Rafael Reef on the east edge of the San Rafael Swell (see map page 20). (Several of the monoclines were called "reefs" by pioneers who likened the rugged topography to oceanic reefs that hindered movement of their prairie "schooners.") All are interpreted to be sedimentary layers draped over deep-seated faults that originated in Precambrian time and were pushed eastward by forces generated much later, during the Laramide Orogeny.

A large monoclinal uplift, the Monument Upwarp, extends from Kayenta, Arizona, some 90 miles north to the confluence of the Green and Colorado rivers in Canyonlands National Park. Lake Powell is entrenched in Glen Canyon along the gently dipping west flank of the large fold, and the Comb Ridge monocline forms its sharply folded eastern margin. Numerous smaller anticlines and synclines are perched on the Upwarp; all were relatively insignificant until the Laramide Orogeny enhanced the structures between 60 and 70 million years ago. Sedimentary rock layers are tilted gently toward the north for many miles along the plunging nose of the Monument Upwarp, greatly affecting the appearance of the landscape in northern Canyonlands National Park.

BASINS

Basins are somewhat like synclines, only much larger. Uplifts and basins go hand in hand, more or less alternating across the province. The Monument Upwarp, for example, is flanked by the Henry basin on the west and the Blanding basin on the east. Basins, like uplifts, usually have smaller anticlines and synclines within the overall structure.

Erosional processes that eventually put the "cosmetic touches" to a

Block diagrams illustrating the three kinds of folds found in layered rocks. Block A is a diagrammatic view of a monocline, a fold that forms where layers of sedimentary rocks drape across a deep-seated fault. Block B represents an anticline (up-fold) and an adjacent syncline (down-fold). Folds such as these form where the Earth's crust has been compressed, or squeezed, from right to left in the figure.

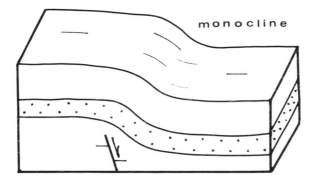

A

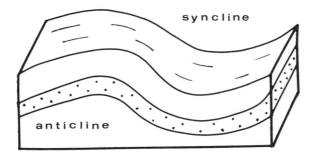

B

landscape are always much more active at high elevations than in lowlands. The uplifts are consequently more deeply eroded than the basins, resulting in the oldest rocks only being exposed on the uplifts and in the deepest canyons. The youngest rocks occur at the surfaces of the basins, where older rocks are deeply buried.

In some cases, however, a whole basin may be so deeply buried that it is unrecognizable at the surface. One such case is the Paradox basin, an ancient sedimentary basin that formed some 300 million years ago in southeastern Utah and southwestern Colorado. It has subsequently been buried from view by thousands of feet of younger deposits. Without deep exploratory drill holes, its extent and nature would be unknown. Thick salt beds were deposited within the Paradox basin, and now lie deep beneath the surface of Canyonlands Country. Though the salt is not visible, its presence is of immense importance in shaping the present-day landscape of Canyonlands Country, and will be discussed in greater detail in Chapter Six.

IGNEOUS INTRUSIONS

Another characteristic feature of the Colorado Plateau Province is the presence of so-called "laccolithic mountain ranges," isolated clumps of high, often snow-capped mountains, that stand as major landmarks or geographic beacons against the otherwise plateau-like landscape. These prominent ranges may seem to be randomly scattered across the Plateau, but in fact appear to be located above major faults in the basement rock or at intersections of deep-seated faults, natural avenues for the vertical movement of molten masses. The word "igneous" refers to fire, or great heat, thus describing the origins of these rock bodies.

The ranges formed as molten rock (magma) flowed upward through the basement faults and gobbled its way upward into the sedimentary layers, melting some, pushing others apart, or breaking through still others. If the columns of magma had reached the surface, lavas would have poured out to form large volcanoes, but the magma cooled and turned into rock beneath the surface. When this happens, much more time is required for the magma to cool and form rock. Consequently, rocks formed deep underground are coarsely crystalline, as in granite. These "intrusive igneous bodies" have subsequently been exhumed by erosion, exposing their cores of strangely shaped bodies of hard rock.

The term "laccolith" implies that the igneous bodies are mushroom-shaped, having a more or less centrally located vertical feeder neck from which emanates a generally circular, lens-like bulged sheet of intrusive rocks. (The word "laccolith" is derived from the Greek term for a "cistern" or "pool." In this case it is a pooling place for magma, although in an underground, domed-up "cistern" formed by forced entry of the molten rock from beneath.) This characterization conjures up images of domed mountains, which the ranges generally appear to be, but it is an incomplete description of the actual shapes of the igneous bodies. In reality, the intrusive bodies are much more complex in shape, resembling staghorn cacti more than mushrooms. In his classic study of the Henry Mountains, Charlie Hunt coined the term "cactolith," far more descriptive of their actual shapes: "a quasi-horizontal chonolith composed of anastomosing ductoliths whose distal ends curl like a harpolith, thin like a sphenolith, or bulge discordantly like an akmolith or ethmolith." (And any fool knows what a thing like that could do.) What all that gobbledy-gook says is that the intrusive rock bodies are "cactus-shaped," but when one looks up the meaning of the individual terms, the shape becomes an impossibility.

The prominent mountain ranges punctuate the plateau country in stately fashion. They include the Henry, La Sal, Abajo, Ute, La Plata and Carrizo ranges, and Navajo Mountain. Their locations are shown on the following map.

SO WHAT?

The Colorado Plateau Province is a region of damnable incongruities: it is either too hot or too cold; either too dry or too wet; depending on the roads you travel, it either is too flat and dismal or too precipitous and colorful; it is treacherous country or a wonderland, depending on one's attitude; the geology is too simple or impossibly difficult to understand, depending again on one's point of view.

Compared with adjacent provinces, the Colorado Plateau is geologically simple, but the details of its individual features have been difficult to understand. For decades, the obvious surface geologic features were considered to have originated during the Laramide Orogeny, a time of mountain building that occurred between 60 and 70 million years ago. Laramide compression came in waves from west to east, buckling the crust accordingly. But why, then, are the major structures oriented in such peculiar disorder? Why do some folds trend north-south, and others east-west? Why has the displacement of rocks along faults been "backwards" to what is "expected"? And why are some of the largest folds oriented obliquely to the "known" compressional forces? These, and a multitude of other problems, have plagued geologists for generations.

The simple fact is that the structures were not formed by Laramide forces, but that the west-to-east push merely modified and finalized them. Preexisting folds and faults had been present since the Precambrian, "set in concrete" more than a billion years before the Laramide Orogeny was even a gleam in anyone's eye. The Laramide event only did what it could with what it had to work with, and that was already a mess.

Many geologic horror stories spontaneously evaporate when that concept is seriously considered. Innumerable "unexplainable" geologic phenomena become almost too simple when we realize that the highly anointed Laramide Orogeny was merely a geologic "afterthought." Such geologic misconceptions are exactly why we must look back to the beginnings of Earth history to gain an understanding and full appreciation of the country we fondly know as Canyonlands.

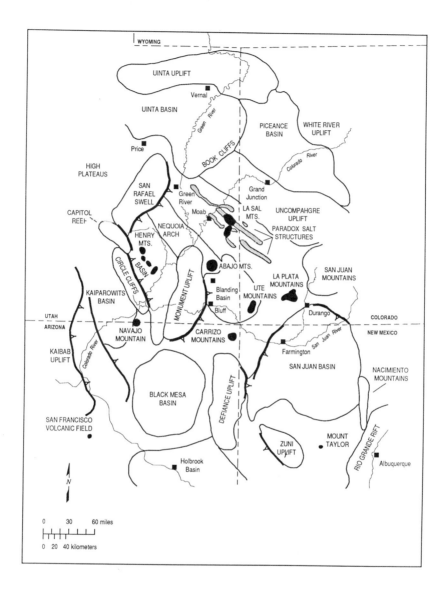

Map showing the major geological structures of the eastern Colorado Plateau Province. Heavy lines with barbs are the major Monoclines that form the margins of some uplifts and adjacent basin; barbs point to the down-side of the fold. Solid black areas represent the intrusive igneous mountains and stippled patterns are salt-intruded anticlines near Moab.

CHAPTER THREE
IT'S ABOUT TIME...

Few ideas are more difficult for the human mind to comprehend than the concept of geologic time. We tend to think of time in terms of a few days, weeks, or years, even a few decades, for these constitute a lifetime. If we think of a few hundred years' time, it is considered to be ancient history. It is impossible for us to conceive of the number of Monday Night Football games we would miss seeing in a million years, a hundred million years, or (heaven help us!) a billion years. Yet the geologic history of Canyonlands Country goes back more than 1.8 billion years!

Geologists are no better than ordinary folks in this matter. They generally are not good mathematicians anyway, and don't really want to fool around with a lot of numbers. Old time geologists were no different in this respect, and so they devised the Geologic Time Scale. It was a scheme whereby they could put chunks of time into handy boxes for easier reference and name the time boxes for pretty places where they liked to look at particular layers of rocks. Each box, a geologic time "period," represented the time it took to deposit one's favorite bed of layered rocks - a "system" - whatever length of time that might be.

Early attempts to classify spans of time were not very successful. For example, Johann Gottlieb Lehmann proposed, in 1766, that times of mountain-building be classified into (1) those that formed at the time of the creation of the Earth, (2) those that formed during the time of "The Flood," and (3) those that formed since "The Flood." However, there have been hundreds of "Floods" during the course of geologic time, so that scheme didn't work well.

By the 1830s, the scheme we now use began to emerge from the work of a few British geologists, but not without some frustration. Adam Sedgwick studied and published descriptions of the oldest sedimentary rock sequence in the pleasant countryside in Wales, naming it the Cambrian System, "Cambria" being the Latin name for Wales. Essentially simultaneously, Sir Roderick Murchison studied the sedimentary rocks along the border between England and Wales, naming his sequence the Silurian System, the "Silures" having been early inhabitants of the region. Both names were published in 1835.

As luck would have it, some of the rocks included in each "system" were the same strata; the two "systems" overlapped. Of course, a professional skirmish broke out. It continued until 1879, when Charles Lapworth settled the matter by naming the overlapping section the Ordovician System for another ancient tribe of people, the "Ordovices." Meanwhile, Murchison and Sedgwick, in 1839, named the next younger geologic period the "Devonian" for Devonshire, a lovely summer resort area in southwestern

England. (Occasional reference to the charts inside either cover of this book may help to mentally organize these names.)

The next younger sequence of rocks in the region is the "Carboniferous System," the "coal measures" of central England. The actual coal-bearing beds are in the upper part, overlying a massive cliff-forming limestone known throughout western Europe as the Mountain Limestone. Although of totally different origin and surface appearance, the two rock types are still called the "Lower" and "Upper" Carboniferous in Europe. An almost identical sequence occurs in the eastern United States as well, where it was subdivided into the older (and therefore lower) "Mississippian System" and the younger "Pennsylvanian System" in 1869 and 1891 respectively. We use the two names effectively in this country, but European geologists still refuse to recognize the only American names in the Geologic Time Scale.

Sir Roderick Murchison became obsessed with naming all rocks for use in an international Geologic Time Scale. He knew that an unnamed series of rocks existed between the Carboniferous System of England and the much younger Triassic System of western Europe, but he was at a loss for an appropriate name. He somehow wangled an invitation from Czar Nicholas I to visit Russia during the summers of 1840 and 1841 under the guise of determining if his other named "systems" of England and Wales were useful in eastern Europe. He traveled extensively into the Ural Mountains, where he found the missing section rather well exposed. In an 1841 report to the czar, he named the section of mostly red beds and gypsum the "Permian System" for the city of Perm and the extensive Perm Basin. Unfortunately, the section Murchison studied contains few distinctive fossils, and the lower boundary of the Permian System has never been acceptable on an international basis.

At first, the "absolute" ages of rocks were an unknown, and the named boxes of geologic time were shuffled into order based on their relative position in the overall sequence of rocks. Thus, the oldest rocks were considered to be those at the bottom of the stack, as they must have already been in place before the next overlying layer could be deposited on top. This idea has become known as the "law of superposition." During the ensuing 150 years, methods of measuring the age of rocks in years, at least in a loose way, have been developed.

The seven named geologic "periods" we have discussed are lumped into a larger unit, the "Paleozoic Era" (from the Greek palaios meaning "old" and zoe meaning "life," or "ancient life"). Paleozoic time was preceded by the "Precambrian Era," the first four billion years or so of Earth history before Sedgwick's Cambrian System was deposited. No universally acceptable subdivision of the very lengthy Precambrian Era has ever been proposed despite numerous attempts. The Paleozoic Era was followed by the Mesozoic ("middle life") and that, in turn, by the Cenozoic ("late life").

The Mesozoic Era is subdivided into three geologic periods: the Triassic Period, named for a tripartite sequence of rocks in Germany; the Jurassic Period, named for rocks of the Jura Alps along the French-Swiss border; and the Cretaceous Period, named for the chalk cliffs bordering the English Channel (from the Latin creta meaning "chalk").

The Cenozoic Era is subdivided into two periods, the names being carryovers from the first known geologic time scale published in 1760 by

Giovanni Arduina. He designated four periods, two of which "stuck" in the present-day time scale: the Tertiary (third) Period and Quaternary (fourth) Period.

The Geologic Time Scale, and the formations deposited in Canyonlands Country during these time periods is printed inside the covers of this book for convenient reference.

CHAPTER FOUR
IN THE BEGINNING...

No one really knows much about the earliest history of the Earth, the Colorado Plateau, or Canyonlands Country. It is truly "lost in antiquity." We think the Earth is about 4.5 billion years old, but the rock record for most of that great time span is lost or so deeply buried that we may never find it. The oldest rocks known anywhere near Canyonlands Country are in the Uncompahgre uplift to the east, the San Juan Mountains of southwestern Colorado, and in the Grand Canyon of northern Arizona, and these date at around 1.8 billion years. By that time, the rocks had been pretty well mutilated and destroyed - geologists say metamorphosed - leaving few recognizable clues regarding their origins. These highly baked and stirred and generally mangled gneisses, schists and granitic rocks have not changed much since those early days. Such intensive global metamorphism never occurred again.

By the time the first sedimentary layers were deposited on the very ancient metamorphosed basement, erosion had nearly leveled the surface of the continent. The record of the missing time at that erosional surface is not well known, but it must represent several hundred million years. Such breaks in the rock record are called "unconformities." The major structural weaknesses in the Earth's crust may have been present during that great era of metamorphism, but it has not been proven.

Sometime after 1.8 billion years ago, but still in the Precambrian Era, sedimentary rocks were deposited on the global unconformity on what would become the Colorado Plateau. The shape and configuration of the depositional basin or basins is unknown, but several thousand feet of slightly altered sandstone (quartzite) is present in both the Uinta and San Juan mountains. About 13,000 feet of sandstone and shale of this age are present in Grand Canyon.

By 1.5 billion years ago, long before Cambrian time, a continental-scale swarm of wrench faults cut northwestward through Canyonlands Country. A similar set of wrench faults that trend northeastward intersected the first set in eastern Canyonlands in the Moab area. The northwesterly faults underlie each of the present-day salt valleys, such as Moab, Castle, Fisher, Lisbon, Gypsum, Paradox, Salt, Pine, and Sinbad valleys. The most important of the northeasterly faults generally underlies the course of the Colorado River from Grand Canyon to Grand Junction, Colorado. The faults jostled the land much like the modern San Andreas fault system is doing today in southern California, but no one seemed to care.

Prior to 570 million years ago, the beginning of Cambrian time, the younger Precambrian sedimentary rocks were tilted by block-faulting in Grand Canyon, and corresponding rocks in the Uinta Mountains of northern Utah were folded into an east-west-trending anticline some 150 miles long. The quartzite sequence in the San Juan Mountains was emplaced by wrench faulting between 1.68 to 1.46 billion years ago. These dates were determined by radioactive decay methods. The wrench faults of the San Juan Mountains are known to extend into eastern Canyonlands Country deep underground.

These deep-seated Precambrian faults set the geologic stage, and will come back to haunt us throughout geologic time.

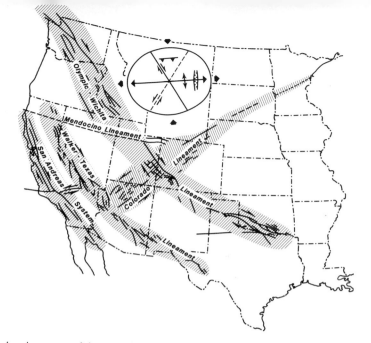

Map showing some of the more important fault zones in the western United States. The Olympic-Wichita Lineament constitutes the northwest-trending swarm of deep-seated faults that underlies the salt-intruded structures of the Paradox basin. Their earliest dated movement was in middle to late Precambrian time. Dating on the northeast-trending Colorado Lineament was about the same. The Walker Lane-Texas Lineament parallels the Olympic-Wichita, but is of a much younger Precambrian age. Faulting along the San Andreas fault system is continuing today, but otherwise it is almost identical to the Precambrian structures.

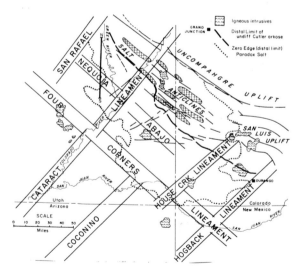

More detailed map showing locations of specific Precambrian and Paleozoic fault zones in the eastern Colorado Plateau. Heavy dotted line shows the outer limits of salt in the Paradox Formation in the Paradox basin.

CHAPTER FIVE
THE EARLY YEARS

About the time life began to flourish in the open oceans of the world during the Cambrian Period, sea level rose on a global basis, at least relative to land levels. Until the beginning of Cambrian time, the sea had been restricted to the region of present-day Nevada and Idaho, but then the beaches marched slowly but doggedly eastward across the coastal lowlands of Utah, Wyoming and Montana. By the close of Cambrian time, some 60 million years later, beach sands were being deposited in eastern Utah and southwestern Colorado. The cliff-forming sandstone that caps the Precambrian rocks in Grand Canyon is called the Tapeats Sandstone, in central Utah it is known as the Tintic Quartzite, and in the San Juan Mountains it is the Ignacio Quartzite. Call it what you wish in Canyonlands; all of these names have been used here.

As the Cambrian shoreline advanced to the east, sea water covered older shorelines, and deeper water, quiet environments replaced the beaches of yesteryear. Mud settled out of suspension to bury the beach sands, and the shells of dead sea life accumulated in the muds to date the passing. The resulting fossiliferous shales are known as the Bright Angel Shale in Grand Canyon and the Ophir Shale in central Utah.

Sea level finally stabilized in Late Cambrian time, and a broad, shallow marine shelf formed where Canyonlands is today. Lime muds were deposited on the shallow shelf, later to "bake" to concretelike limestones in the hot sun, and become the uppermost Cambrian deposits in the region. These rocks are found only in deep wells drilled in the search for oil in Canyonlands.

Not much changed for the next 130 million years on what is now the Colorado Plateau. There are no rocks of Ordovician or Silurian age anywhere in the province, as far as anyone knows. Perhaps some were deposited, but if so, erosion won the battle and none were preserved. The resulting surface of erosion and/or nondeposition became an "unconformity" of wide geographic proportions.

Our structural heritage, the basement faults, were fairly quiet during Cambrian time. There was only minor movement along the faults as compared to the havoc they created in the "good old days."

THE SECOND WAVE

And then it happened again. Late in the Devonian Period, sea level rose to inundate the landscape after a hiatus of nearly three geologic periods. There were no sandy beaches as in the Cambrian Period. Instead, a muddy tidal flat blanketed the region as far east as central Colorado. The resulting rocks, known as the Elbert Formation (see time chart inside the covers), consisted mostly of lime muds and green clays, but a few sand bars developed here and there. They built up in the surf zones where waters shallowed on the fault blocks inherited from the basement.

Sea level continued to rise slowly and lime muds containing some marine fossils were deposited over the Elbert tidal flats. The resulting rocks, the Ouray Limestone, were deposited in very shallow water, perhaps ankle-deep to hip-deep, but they represent the time of maximum water depth. Fossils suggest that deposition may have continued without interruption into earliest Mississippian time, some 360 million years ago.

Sea level subsided once again in Early Mississippian time, and limestones of the Ouray changed gradually into thin-bedded and shaly dolomites (magnesian-rich limestone) of the lower Redwall Limestone, representing tidal flats that again spread across the region. This shallowing trend ended with the sea withdrawing completely from the Colorado Plateau country early in Mississippian time. The resulting erosional surface is present within the Redwall Formation in Grand Canyon and in cores of deep wells drilled in Canyonlands Country, where the formation is called the "Leadville."

Thus, a cycle of sedimentation ended. It started with the tidal flats of the Elbert Formation, changed slowly to deeper-water limestones in the Ouray, and ended with a return of tidal flat conditions in earliest Mississippian time. The cycle was finalized as the tidal flats emerged and dried to a concretelike surface that occurs midway up in the Redwall (or Leadville) Formation.

CRINOIDS INHERIT THE EARTH

As quickly and quietly as the seas left Canyonlands Country, they returned. This time the waters quickly deepened, and fossiliferous marine lime sediments were deposited across the entire western part of the continent and most of the world. The resulting limestones of Mississippian age (Lower Carboniferous to the Europeans) form the upper member of the Redwall (Leadville) Formation in Canyonlands Country. Waters were apparently fairly deep, perhaps 10 or 20 feet deep, because the sediments were lime muds deposited in quiet water. Where the sea shallowed over shoals on slightly rising old basement fault blocks, fossiliferous mud banks developed under ideal conditions for all sorts of sea life to flourish. As it turns out, marine algae (primitive plant life) love the shallowest waters where sunlight is most intense. And everyone in the ocean loves to eat algae or other critters that eat algae. So all sorts of shell life piled up into mud banks under these conditions.

At this stage in the evolution of marine life, "crinoids" were extremely abundant worldwide. They are marine animals that lived attached to the sea floor, the main body of the animal being held well above the bottom sediments on stalks of loosely fitted, button-shaped, lime segments. When the animal later died and fell apart, the skeletal parts became lime sand grains that piled up into mounds that are common on the higher fault blocks. All of the animals shown on the following illustration are common fossils found in these deposits.

Fossils commonly found in rocks of late Paleozoic age. Crinoids, plant-like animals, were especially abundant in limestones of Mississippian age. They grew in thickets, each animal attached to the bottom by hold-fasts, and disagregated into button-shaped segments upon death. Brachiopods are rather abundant in marine sedimentary rocks, having filled ecological niches much like the clams of today. They differ from clams in that each shell is bilaterally symmetrical, rather than being symmetrical along the hingeline between the shells as in clams. Both lived in late Paleozoic seas, but clams are rather rare, while brachiopods are very common in these rocks. There are thousands of species of brachiopods, each having a different appearance. There are also myriad varieties of corals, the one shown here being common in late Paleozoic rocks. These are ancestral to modern corals, but are biologically distinctive and are now extinct. Drawings are by William L. Chesser.

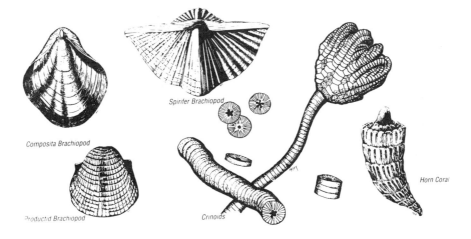

Composita Brachiopod

Spirifer Brachiopod

Productid Brachiopod

Crinoids

Horn Coral

Not too unexpectedly by now, the widespread shallow seas again withdrew from Canyonlands Country back westward to the main seaway. During the later half of the Mississippian Period, the resulting lowlands lay sweltering in a hot, humid, subtropical climate, and red, iron-rich soils - Molas Formation - developed from weathering of the limestone surface. Thus ended a long and relatively peaceful era.

EARLY OIL

Rocks of Devonian and Mississippian age host a few oil fields in the vicinity of Canyonlands Country. The largest and richest of these is the Lisbon Field that lies some 25 miles southeast of Moab, Utah. Another field that produces more gas than oil is the Southeast Lisbon Field, sometimes called McIntyre Canyon, that is another 12 miles to the southeast. The Big Flat Field is in the grassy flats near the junction of the Dead Horse Point and Canyonlands National Park roads northwest of Moab. Finally, the Salt Wash Field is about 25 miles northwest of Big Flat. All occur along a nearly straight line that trends northwest-southeast along one of our basement faults, as shown on the map (page 26). Some of the oil is being produced from ancient offshore sand bars found in the lower Elbert Formation of Devonian age. Most of the oil, however, comes from the crinoid banks in the upper Redwall Limestone of Mississippian age.

It was drilling in search of these and other possible accumulations of oil and gas that has given us the information about these older rocks under Canyonlands Country and the deeply buried fault system that is so intimately involved.

CHAPTER SIX
IT HITS THE FAN

So much for peace and serenity! Sometime around the beginning of Pennsylvanian time, earthquakes started rumbling through what is now southwestern Colorado and spread rapidly into east-central Utah. Our basement faults were acting up again, this time in earnest! We call this the start of the Ancestral Rockies Orogeny in the western United States; it was an Earth-shaking event, to say the least. The faulting had started in Precambrian time, then the faults were relatively dormant for some 1.5 billion years, and all this would eventually lead to the Laramide Orogeny, another 230 million years later.

The term "orogeny" simply means "mountain building" and indeed, **mountains were built** during the Pennsylvanian Period. They arose along fault blocks in Oklahoma and the uplifts spread rapidly into the Ancestral Rocky Mountains of Colorado. Several high ranges were formed: the Front Range Uplift to the east in central Colorado, the Wet Mountains and Sawatch Uplifts in the middle, and the San Luis Uplift in southwestern Colorado and north-central New Mexico. The one that affects us most, the Uncompahgre Uplift, became a prominent mountain range by Middle Pennsylvanian time. All were bounded by major faults that formed highlands and offset the landscape laterally exactly as the San Andreas fault system is doing in California today. Each fault block was moving northwesterly relative to the neighboring landscape to the east.

What in the world would cause this renewed shuffling of huge blocks of the Earth's crust? It was obviously an event of great north-to-south crustal compression (or was it south-to-north?), and east-to-west extension of the Earth's crust resulted. In the process some blocks popped up, as others dropped. Some say it all happened when the South American plate crashed into the North American plate, sparks flying, as a consequence of a plate tectonics-related continental collision.

Adjacent to each mountain range, basins sagged into existence. These deep, down-faulted troughs served as waste disposal sites for thousands of cubic miles of boulders, sand, and dirt that would wash down from the mountains. As will happen in any self-respecting mountain range, weather patterns were altered, and torrential rains and melting snows provided the running water to carry bits of weathered rock toward the lowlands.

It wasn't long - at least in geologic time - until erosion bared the mountains to their cores of Precambrian metamorphic and granitic rocks. The sand and gravel brought to the basins by streams were composed mostly of quartz, feldspar, and iron-bearing minerals such as augite and hornblende, because they were abundant in the parent rocks. These mineral grains, when deposited together, form rocks called "arkose." The iron-bearing minerals easily decompose (basically they "rust") to produce the red colors so obvious in the rocks of Canyonlands Country today. Feldspar grains, the squarish gray or pink grains in granite, do not decompose as quickly, and form the pink grains of sand so characteristic of the Cutler Formation.

Red sandstones and conglomerates that fill each basin of the Ancestral Rockies have unique formation names. In the Denver Basin, east of the Front Range Uplift, they are called the Fountain Formation, in the Central Colorado Trough (Eagle Basin) they are the Maroon Formation, and in southwestern Colorado and eastern Utah they comprise the Cutler Formation. The bright red sandstone/conglomerate cliffs in Richardson Amphitheater and Fisher Towers east of Moab are in the Cutler Formation, a formation that presents a multitude of problems that will be the subject of the next chapter.

SALT

Perhaps the largest basin associated with the Ancestral Rocky Mountains is the Paradox basin, lying southwest of the Uncompahgre Uplift in southwestern Colorado and eastern Utah. It is a long structural depression extending from near the Colorado-New Mexico border northwestward for 200 miles nearly to Green River, Utah. It is a complexly faulted basin that steps down from the southwest toward its deep axis adjacent to the Uncompahgre Uplift.

As already discussed, the Paradox basin contains thousands of feet of arkose (sandstone containing a lot of feldspar grains) derived from metamorphic and granitic rocks of the adjacent highland on its eastern flank. The rest of the basin, away from the mountains, was a deep depression as the sea entered the region from the west and south in Pennsylvanian time. Several land areas intervened as the rising sea found its way into the stagnant inland basin, these being the ancestral counterparts of the present-day Zuni, Defiance, Monument, and San Rafael uplifts. The scenario is shown on the following illustrations.

To further complicate matters, sea level fluctuated many times during Pennsylvanian and Early Permian time, affecting events on a global scale. The cause was probably repetitious glacial advances in the polar regions, much like the "ice ages" of the past million years or so. When polar glaciers accumulate and advance, vast amounts of water are locked up in ice, causing sea level to lower, but when the glaciers melt and retreat, sea level again rises. This is similar to daily tides, but on cycles of millions of years. So sedimentation in the Paradox basin and elsewhere in the world was cyclic.

RECIPE FOR SALT

When a bucket of sea water is left out in the hot sun to evaporate, the minerals in it precipitate from solution. First to crystallize are the carbonates, that is, limestone or dolomite. Second, gypsum (hydrous calcium sulfate) settles in the bucket, and if evaporation continues, salt (sodium chloride) crystallizes from the solution. Finally, if the water completely

32

Block diagram showing the region that would become the Colorado Plateau and Southern Rocky Mountain provinces at about Middle Pennsylvanian time, some 300 million years ago. The main seaway lay to the west, in western Utah and Nevada, bordered on the east by large low-lying areas that at times rose above sea level. The ancient uplifts nearly isolated the Paradox basin from the open sea, and poor water circulation caused great thicknesses of salt to be deposited. Fault-block mountain ranges rose to the east, where rivers carried sediments to the nearly land-locked basins. (Drawing by William L. Chesser, Courtesy of Canyonlands National Park)

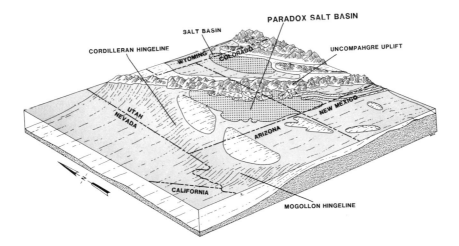

Block diagram of the eastern part of the Colorado Plateau in Middle Pennsylvanian time, about 300 million years ago. The Paradox basin sagged below sea level along large faults, bordered on the east by an up-faulted mountain range, the Uncompahgre Uplift. On the west, the Monument Upwarp was relatively high, standing at times above sea level, and in general diverting the normal flow of marine currents away from the basin. Stagnation of the nearly isolated sea water in the basin caused thick layers of salt to be deposited. Erosion of the rising Uncompahgre Uplift provided mud, sand, and gravel to rivers that carried the sediments into the eastern margin of the basin, where they mixed with the salt on deltas. (Drawing by William L. Chesser, courtesy of Canyonlands National Park)

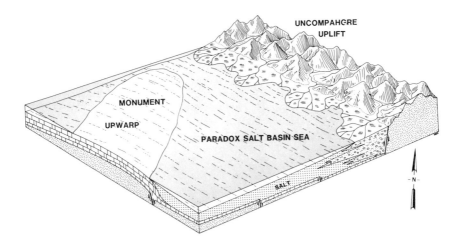

evaporates potash salts will be deposited. Rocks formed in such a natural process are called "evaporites."

And that is exactly how each salt cycle occurs in the Paradox basin. The stagnant basin acts as the bucket, albeit a huge one, and each cycle requires refilling the bucket with fresh sea water. A rock cycle usually begins with a bed of dolomite, followed upward by gypsum, which in turn is overlain by a thick bed of salt. In some cases the sequence is topped off by a bed of potash salt. The cycles are each separated by black shales above and below. Twenty-nine such cycles are recognized in the Paradox Formation, totaling several thousand feet in thickness in some parts of the basin.

How did such thick salt accumulate in the Paradox basin, back some 300 million years ago? The shape of the basin permitted sea water to enter across its shallow margins, but then be trapped and left to evaporate and precipitate salt in the arid climate. A natural conveyor belt continually, or episodically, supplied new salt to the stagnant basin with each rise of sea level.

SQUISHY SALT

Surprising as it may seem, rock salt flows like wet putty when under high confining pressures. For example, if salt is buried deeply under other sedimentary rocks, a great deal of pressure is exerted on it by the weight of the overburden. Under these circumstances, salt flows somewhat like glacial ice, toward any area of lower pressure. Deep underground it can flow laterally, vertically, or both, depending upon the nature of the pressures acting on the salt.

Salt was deposited to great thicknesses as the Paradox basin was actively sagging along major faults. Meanwhile, thousands of cubic miles of gravel, sand, and dirt were being carried down from the Uncompahgre Uplift, filling the eastern margin of the basin. The weight of sediments being dumped onto the salt from the east created a heavy, lopsided overburden. As a result, the salt began flowing away from the overburden, generally toward the southwest. When the mobile salt encountered the large fault blocks already present, it was deflected upward. Overlying layers of rock were arched at first, but later in the process they were actually pierced by the salt, forming structures called "diapirs," not to be confused with diapers.

Topographic bulges developed at the surface where salt flowage lifted the overlying strata, and synclines were formed by salt withdrawal along the flanks of the bulges. Streams flowing from the uplands were diverted around the hilly terrain, and their sediments were deposited in the synclinal valleys. Thus, differential pressures caused by varying thicknesses of overburden were locally exaggerated, causing salt flowage to continue from Pennsylvanian time through the Jurassic Period, some 150 million years.

Salt flowage ended when the mother lode of bedded salt was depleted in adjacent synclines, and the structures were given a deep burial by thick marine sediments of Cretaceous age. Salt is again rising in some of the structures, but now it is caused by erosion of those overlying sedimentary rocks, removing the weight of overburden. The salt in the cores of the structures is simply rebounding as the overburden is relieved.

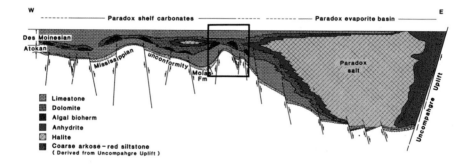

W   ------- Paradox shelf carbonates -------   -------   Paradox evaporite basin -------   E

Des Moinesian
Atokan
Mississippian   unconformity   Mola Fm   Paradox salt   Uncompahgre Uplift

Limestone
Dolomite
Algal bioherm
Anhydrite
Halite
Coarse arkose – red siltstone
(Derived from Uncompahgre Uplift)

Diagrammatic cross section across the southwestern shelf of the Paradox basin and the deeper, salt-filled eastern basin to the Uncompahgre Uplift on the right. The basin dropped down along numerous faults and the stagnated basin filled with evaporites in Middle Pennsylvanian time. At about the same time, the shallower shelf areas accumulated limy sediments, and algal banks that now contain prolific oil deposits were built on the fault-controlled shoals.

The faults that triggered the upward flow of salt are not small features. Drilling at Paradox Valley found that the fault escarpment beneath the salt structure has a vertical relief of more than a mile. A well drilled on the southwest flank found zero salt, but a hole drilled northwest of the fault penetrated nearly 15,000 feet of salt; rocks beneath the salt were down-faulted about 6,000 feet lower than in the southwest flank well. The relationships are shown diagrammatically on the following illustration.

There are numerous salt-flowage structures in the eastern Paradox basin where the original bedded salt was thickest. They are all northwest-trending structures that form long, narrow valleys at the surface today. The largest of these are Gypsum, Paradox, Pine Ridge, Moab, Castle, Fisher, Sinbad, and Salt valleys. All have major faults underlying their southwest flanks; all have thousands of feet of salt under the valley floors. They are present-day valleys because groundwater has dissolved salt from the tops of the flowed salt bodies, and overlying layers have collapsed into the voids left by salt dissolution.

OIL

As previously mentioned, the cycles of evaporitic rocks are separated by black shale beds that extend throughout the Paradox basin. They are black because they contain large amounts of organic material that was changed to oil by considerable heat and pressure over the eons. The black shales were the source rocks for untold jillions of barrels of oil, some found and some unfound, in rocks of the Paradox basin. However, little oil is produced from these rocks in Canyonlands Country because they are too fine-grained to permit the flow of oil. But it is different elsewhere!

The shallow margins of the Paradox basin were teeming with marine life in the clear, warm sea water back in Middle Pennsylvanian time, some 300 million years ago. The shallow shelf environments ringed the deeper evaporite basin to the south and west of Canyonlands Country. Critters that secreted limy shell material were responsible for the sediments that produced limestones that are roughly equivalent in age to salt and gypsum beds of the deeper basin.

A special kind of lime-secreting algae flourished in the shallow environments, sometimes to the near exclusion of the usually abundant animal life. These green, marine plants secreted calcium carbonate within their tissues that hardened when the plants died, forming fossil fragments that piled up like corn flakes on the sea floor. Thick accumulations of the algal debris built mounds of lime sediments on fault-bounded shoals scattered across the shallow platform. Not only are the mounds numerous in any one cycle, they recur repeatedly in several cycles. Oil was "squeezed" from the black shale source beds into the porous and permeable sediment banks (reservoir rocks), especially in the Blanding Basin south of Canyonlands Country.

The biggest and best of these oil fields is the Aneth Field in southeastern Utah. It will produce more than 500 million barrels of oil before it is depleted, making it one of the nation's "giant" oil fields. There are dozens of other smaller, but very lucrative, oil fields in the Blanding Basin.

Diagrammatic cross section as if the Earth were cut with a knife across Lisbon and Paradox valleys to the Uncompahgre Uplift on the right (NE). The patterned pillar-like features are salt-intruded anticlines (diapirs), the one to the left represents Lisbon Valley and that to the right is Paradox Valley. Major basement faults underlie the southwest (left) flank of each structure. The nature of the rocks underlying the Paradox salt changes abruptly near the faults, indicating the faults were actively affecting the sea floor and water depths as the sediments were being deposited. Holes drilled along the southwest (left) flank of Paradox Valley encountered no salt and no Leadville (Mississippian) rocks, but a well drilled in the center of the valley penetrated nearly 15,000 feet of salt and normal Leadville under the valley as shown by the vertical lines. The intervening fault has more than a mile of vertical displacement. Bedded salt began to flow deep underground toward the southwest (right to left) in Middle Pennsylvanian time, away from thick accumulations of sand and gravel that were dumped into the eastern (right side) of the basin from the high Uncompahgre Uplift. Where the flowing salt encountered the large faults, it was deflected upward and eventually pierced all overlying rock layers. Solution of the near-surface salt by ground water caused overlying rocks to collapse back into the tops of the structures during the past couple of million years. As indicated by the oil well symbol, oil and natural gas are being produced from Leadville and lower Elbert rocks along the southwest flank of Lisbon Valley. Symbols along the left margin of the cartoon abbreviate the geologic periods.

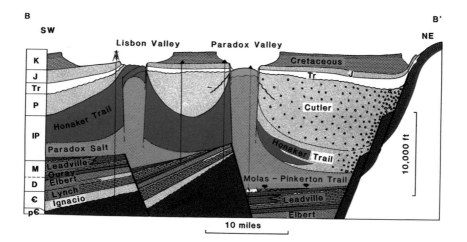

Shafer dome from the Shafer Trail-Potash road. Note that the rocks forming the light-colored benches in the middle distance are dipping gently, both to the right and left of the center of the view. This is an anticline, in this case pushed upward by salt flowage at depth. The light-colored, ledge-forming bed is here known as the "Shafer lime."

PARADOX?

There are many paradoxical aspects in the Paradox basin, but the name stems from a specific feature. Early pioneers visiting the region noticed that the Dolores River crosses a large valley, rather than flowing along the valley as would be expected. They realized that something was really screwy, and so they named it Paradox Valley.

Then geologists of the U.S. Geological Survey mapped peculiar exposures of gypsum and black shale in the valley floor in the 1920s. They didn't understand the occurrence, but realized the rocks were unusual. Consequently, they named the evaporites the Paradox Formation for their occurrence in Paradox Valley. Later, as drilling for oil proceeded, the age and geographic extent of the evaporites were realized, and the depositional basin became known as the Paradox basin.

OPEN SEAS: HONAKER TRAIL FORMATION

By the end of Middle Pennsylvanian time, salt sedimentation had run its course. The salt basin gradually shrank and retreated farther to the northwest with each ensuing depositional cycle. The Paradox basin was literally filled to capacity with salt and other evaporites, as normal marine environments gradually returned to the region. Sedimentation was still cyclic, but by Late Pennsylvanian time the cycles consisted of alternating fossiliferous limestones and marine sandstones and shales. Shallow seas again dominated Canyonlands Country.

Arkosic sands and gravels were still washing into the eastern margin of the seaway from the Uncompahgre Uplift. It is a good bet that much of the several thousand feet of coarse sandstone is Late Pennsylvanian in age, but because no diagnostic fossils have been found there is no way to be sure. These stream and deltaic deposits occur east of Moab in and near the salt structures, where few marine limestones are interbedded with the stream sands.

West of the salt structures, marine limestones dominate the sequence of Late Pennsylvanian rocks. The entire section, known as the Honaker Trail Formation, is about a thousand feet thick. It was named for the trail into the San Juan River canyon west of The Goosenecks. The highly fossiliferous rocks make up most of the gray, ledgy cliffs in Cataract Canyon, from the confluence of the Green and Colorado rivers downstream nearly to Mille Crag Bend. Fossils include abundant brachiopods, bryozoa, corals, and crinoids, a few clams and snails, and rare trilobite fragments.

PENNSYLVANIAN-PERMIAN BOUNDARY

Finding the layer of rock that represents the end of Pennsylvanian time can be troublesome in Canyonlands Country. Just when you think you have found it, some fool geologist will argue about it. In any one outcrop, the contact between the geologic periods may be impossible to find; fortunately, the dilemma is easier to solve on a regional basis.

Cataract Canyon at the mouth of an unnamed tributary down-river from Tilted Park on the Colorado River. The steeply dipping rocks in the foreground are gypsum that has been squeezed upward from the Paradox Formation at shallow depth to form a "Gypsum plug" on the Meander anticline. The canyon walls are highly shattered and slumped, giving them a very unorganized appearance.

The problem lies in finding and correctly identifying the right kinds of fossils. Those that work best are microfossils, single-celled animals known as "fusulinids," but they can be difficult to locate in the rocks. They are about the size and shape of a grain of wheat, and extremely complex on the inside. In fact, they look a lot like microscopic jelly rolls, with sheets of chambers rolled up into spindle-shaped coils. To see the details, one must slice them in half, making sheets of rock so thin that light can pass through. Then a genus and species name can be determined by studying the structure under a microscope. Each species was very short lived, so when a fossil is identified correctly, the time in which it lived can be determined rather accurately. But all this takes the skill of a brain surgeon and the patience of Job. And as with any religion, one must believe, above all! After fifty years of study, there is still disagreement regarding the position of the Pennsylvanian-Permian boundary in Canyonlands Country.

Streams carried sand and gravel from the Uncompahgre highlands into the eastern Paradox basin, continuing without apparent interruption or change into Permian time. The paucity of marine fossils in the rocks makes it impossible to find the time boundary in this region. Sediments filled the lowlands, and made a futile attempt to bury the growing salt structures. Gradually and surely the great influx of sediment deposited in vast deltas crowded the sea from the area, never to return in the same form again.

West of Moab, fossiliferous limestones of Permian age are present in the red sandstones and mudstones, and where they lie on rocks of Pennsylvanian age an erosional surface (unconformity) is present. As the seas withdrew from the region in latest Pennsylvanian time, erosion stripped the lands of varying thicknesses of sedimentary rock. The entire Pennsylvanian section, perhaps several hundred feet thick, was removed by erosion before Permian time on the San Rafael Swell to the west of Canyonlands. Now rocks of Permian age rest directly on Mississippian limestones beneath much of the western Colorado Plateau. Although the exact age of the unconformity is still being argued, its presence in the rocks is significant, marking the end of an era in geologic history dominated by the Paradox basin. Anyway, what would geologists do if there were no good arguments?

Long- and cross-sections cut through a fusulinid, much enlarged.

CHAPTER SEVEN
RED BEDS INHERIT THE EARTH

Canyonlands Country is synonymous with red rocks. Most of these rocks are of Permian age. When we enjoy the gorgeous views from any of the high vantage points (Dead Horse Point, Grand View Point, Green River Overlook, Anticline Overlook, Needles Overlook, or North Point), we are standing above great cliffs of sandstone that are Jurassic in age. The scenes below, however, are of Permian-age strata, extending almost to the horizon. Why then does the landscape below look so different from each of the overlooks? It is because of great complexities in the distribution of various rock types within the Permian System. The nature of the rocks changes rapidly and radically within the confines of Canyonlands National Park, even though the rocks are all nearly the same age.

As in Pennsylvanian time, the Permian Period was a time of fluctuating sea level on a global scale, and Canyonlands Country was no exception. Streams from the Uncompahgre highlands were flooding the coastal plains with sediments, perhaps with more vigor than ever before. However, the salt structures were actively growing, and the resulting topography trapped the coarser sediments in the low valleys between and east of the structures. Meanwhile, the sea had returned, at least to the northwestern corner of what had been the Paradox basin. Fluctuating sea level means wandering shorelines, and when sand is plentiful, migrating beaches and coastal dunes result. The more sea level varies, the more the different but related environs wander. And so it was in Canyonlands Country.

THE SEA

As in the past, the main Permian seaway lay to the west of the Colorado Plateau in what is now western Utah and Nevada. The thickest sedimentary rocks occur in central Utah's Oquirrh (pronounced "**oak**-er") Basin, where marine deposits are more than three miles thick. An embayment of the Oquirrh sea sneaked into Canyonlands Country, via the Provo-Price-Green River route, reaching as far south as Cataract Canyon and as far east as Moab.

The best and most complete exposures of the Lower Permian marine strata in Canyonlands Country occur near the confluence of the Green and Colorado rivers. These interbedded limestones, sandstones, and shales were named the Elephant Canyon Formation for the tributary canyon to the Colorado River three miles above the confluence. There, the bottom of the formation is an angular unconformity near river level, and the top is at the base of the massive, light-colored cliffs of Cedar Mesa Sandstone near the rim of the canyon. Thus, a thousand feet of the gray, ledgy cliffs of the canyon walls at the confluence are in the Elephant Canyon Formation.

Almost all limestone beds of the Elephant Canyon Formation contain fossils of marine invertebrates. Brachiopods, bryozoa, and crinoids are most common, but corals and trilobites are occasionally found. Clams and snails are especially abundant in limestone and shale beds in exposures near the potash mine west of Moab. The formation is dated as Lower Permian on the basis of its fusulinids (microfossils), and the trilobites concur; some geologists do not.

FACIES CHANGE

From the magnificent viewpoints overlooking Canyonlands Country, one observes that the rocks are bedded. It seems like the eye can follow any particular bed until it disappears over the horizon or goes underground. Yet no layer of sedimentary rock extends around the world; each has its geographic limits somewhere.

Imagine that you are standing on the beach along the Texas Gulf Coast; perhaps you are on Padre Island near Corpus Christi. You are looking to the south, out to sea into the Gulf of Mexico, where sand and mud are being washed around in the ocean to be deposited as marine sediments. They will contain the shells of marine animals and other evidence of their marine heritage. The sediment is mostly mud, but there is an offshore sandbar right out there where the waves are breaking. And then there is the sand on the beach where you are standing. This is the edge of the marine sediments now. Turn around and you will be looking at dunes of windblown sand; they are not marine deposits, but "eolian" (windblown) instead. Yet the marine sediments are being deposited at the same time as the dune sand. In the rocks that will inevitably result, there will be a "facies change": marine mud (shale) will pass laterally into beach sand (sandstone with low-angle cross beds) that in turn will change to dune sands (sandstone with large, steep cross bedding), all of exactly the same age.

That's not all. Padre Island is a barrier bar, a sand bar that has built above normal high tide, and there is a lagoon beyond. It contains salty water, but salinities are low because it is being fed by fresh-water streams. Plants and animals live in the lagoon that could not survive in the open ocean, so the resulting fossils will be different. And beyond, streams flow down to the lagoon, carrying and depositing "fluvial" sediments.

Now we are seeing marine mud changing laterally to beach sand, then dune sand that in turn changes to lagoonal mud and eventually to stream deposits - all within sight - all happening at the same time. Lateral changes of this kind, when found in sedimentary rocks, are known as "facies changes."

Notice how thoroughly flat the coastal plain is here on the Texas Gulf Coast. What would happen if sea level were to rise by, say, 10 feet? The shoreline pattern of sedimentation would migrate inland for several miles. Padre Island sand would be buried under marine muds and the shoreline deposits would be inland perhaps five or 10 miles, lying on today's stream deposits. But what if, heaven forbid, sea level were to rise a 100 feet?

This scenario was exactly what happened in Canyonlands Country during Permian time. Sea level rose a few feet and coastal conditions migrated inland for several miles. Then, as sea level dropped, the shoreline moved back toward the sea, and associated sediments followed the shoreline accordingly. Extremely complex facies changes resulted, causing the drastic differences in rock types we see today. Permian rocks in Canyonlands Country are a madhouse of facies changes.

ELEPHANT CANYON FORMATION

As one travels down the Colorado River from Moab toward the confluence, the rocks along the river change dramatically. Limestone beds are at first thin, containing mostly clam and snail fossils, and the red sandstones are thick, comprising most of the canyon walls. As we move on, the limestones become thicker and contain more brachiopods and crinoids, and the red stream and dune sands become progressively thinner. Finally, as we approach the confluence, limestones dominate the canyon walls, and the red beds have nearly disappeared.

In gross form, this is a giant facies change. Each bed of limestone represents a time of maximum extent of the Permian sea; each red sandstone records a lower sea level when continental conditions prevailed. The interfingering of rock types resulted from the cyclic rise and fall of sea level. Coastal lowland environments dominated to the east (Cutler Formation), while marine conditions prevailed near the confluence (Elephant Canyon Formation).

ENTER WHITE

Then another variable entered the scene. Silently and surely white, fine-grained sand appeared, coming seemingly from nowhere. Some would say it was blown into Canyonlands Country on the prevailing northwesterly winds; others that it washed in across the Permian sea. Whatever the truth may be, white sandstone, the Cedar Mesa Sandstone, came to dominate the rock record. As with limestones of the Elephant Canyon Formation, the white Cedar Mesa Sandstone interfingers with the red Cutler arkosic sandstone from the Uncompahgre source area to the east. Beds of white sandstone thin and pinch out eastward into the dominantly red Cutler rocks, while tongues of the red sandstone thin and pinch out westward into the white rock, all within the Needles District of Canyonlands National Park. There, the rock in the cliffs and pinnacles and crazy-shaped knobs is banded red and white. This is the heart of the great red-to-white sandstone facies change. The color alterations may be seen in their true light from a low-flying airplane. The entire section of rock is red along the east side of the Needles District in the vicinity of the Six Shooter Peaks and Dugout Ranch, but entirely white in the vicinity of Cataract Canyon to the west. The two rock types, busily intertonguing in the Needles, represent one of the prettiest facies changes of all!

But just where did the white sand come from and how was it

An ancient tidal channel cut into red sediments (dark layers) of the lower Cutler Formation, later filled with white sand of the Cedar Mesa Sandstone. This exposure is along the Green River in Stillwater Canyon near Turk's Head.

deposited here in Canyonlands Country. Let's try to figure it out.

DUNES

So you think you know what a dune is? It is a pile of sand blown into regular shapes by the wind, right? Yes, but in the vernacular of geologists, dunes are piles of sand formed by currents, be they wind or water. When dunes are cut open and studied internally, they are found to consist of sloping beds of layered sand. These "cross strata" or "cross beds" are formed by sand grains blowing or washing over the dune, being deposited on the downwind, or lee side. The lee slope forms at the "angle of repose" of the sand, meaning the angle at which the slope will stand by itself without slumping. That angle is different in water than in air. Also, the cross beds dip, or slope, in the direction the wind was blowing; after all, they are deposited on the lee side of the dune. Ripples formed on the surface of the sand are quite different in wind and water deposits. These differences would be very noticeable back on Padre Island.

The Cedar Mesa Sandstone is a peculiar body of rock for many reasons. It is composed of numerous thick sets of cross beds that generally appear to be preserved windblown dunes. The angle of dip of the cross beds doesn't match the angle of repose of sand in air, but rather approximates the angle of repose in water. The cross beds invariably dip in almost identical directions - toward the southeast, although windblown sands are usually much more variable. There are darned few ripple marks in the rocks, but those present are water-formed ripples. In many places, the bedding is highly contorted as if it had been balled-up or enrolled, a feature that can only form by slumping in very wet sand. Many of the sand grains are not quartz at all, but when viewed under a microscope are found to be tiny broken fossil fragments of marine animals. And the sandstone often contains grains of the mineral glauconite, a green, amorphous complex iron silicate that forms only in normal seawater.

So one begins to think - maybe the Cedar Mesa Sandstone is of marine origin. It is almost impossible, however, to convince most geologists that such large-scale cross bedding could form under water, so they make up excuses for the peculiarities of the sandstone. The reason for the low angles of dip on the cross beds is that the upper, steeper parts of the lee slopes were not preserved. Anyway, fossil fragments and glauconite grains can be blown around, too, even though they are much softer than ordinary quartz-sand grains. And so on.

One of the biggest hangups in solving this problem is that studying modern windblown dunes has become quite fashionable. Everyone is doing it these days, probably because it provides an excuse to travel to all sorts of exotic places around the world. So we have come to know everything there is to know about windblown dunes. And somewhere in the world there are windblown deposits with every imaginable feature that sand can possibly possess. Thousands of technical reports have been written on the internal and external characteristics of windblown dunes.

On the other hand, no one has yet found a way to cut open an underwater sand bar or dune, so there are no technical papers in print regarding internal bedding features in offshore bars. One might suspect that they would have many features in common with their windblown counterparts. The peculiar crescent-shaped windblown dunes found on any sandy desert are called "barchan" dunes (pronounced "barkin'"). Certainly they must be uniquely typical of windblown dunes. Yet there are thousands of underwater barchan dunes on the Great Bahama Banks!

So who knows? In case you have not figured it out, there is still one bigoted old geologist who thinks much of the Cedar Mesa Sandstone is waterlaid, and probably marine.

THE MARINE CONNECTION

There is other evidence tying the Cedar Mesa Sandstone to marine environments. Not only does the Cedar Mesa interfinger with stream-deposited red beds in the Needles District, it also interfingers with marine limestones of the Elephant Canyon Formation in Cataract Canyon. In fact, the entire Cedar Mesa Sandstone interfingers with the marine Elephant Canyon in the subsurface northwest of the confluence of the Green and Colorado rivers. This revelation places the Cedar Mesa Sandstone at, or very near, the scene of the crime. If indeed the sandstone is entirely of windblown origin, it all happened within sight of the ocean. Some of it is probably marine and other parts are windblown in a "coastal" environment.

Whatever its origin, the Cedar Mesa Sandstone is a huge body of sand. The formation thickens to more than 1,200 feet near Hite in southwestern Canyonlands. It extends through Grand Canyon, where it is called the Esplanade Sandstone of the Supai Group, and west into Nevada.

But where did all that white sand come from? The strongly oriented cross bedding in the Cedar Mesa Sandstone indicates that wind and wind-driven currents were coming from the northwest. This suggests that a source area existed somewhere in that direction, but the nearest possibility would be in northern Idaho and western Montana. Anyway, the sand had to be carried more than a hundred miles across the Elephant Canyon sea by wind-driven currents or it had to have been entirely airborne, and that seems unlikely. So the source of the sand remains a mystery.

AND MORE RED BEDS

When the supply of white sand dwindled and Cedar Mesa time ended, as all good things must, red sediments from the Uncompahgre Uplift again dominated the scene. This time the red muds and silts were widely distributed by streams, the coarser sands being deposited near the highlands, while the finer material reached coastal lowlands and tidal flats as far west as Nevada. The resulting soft-weathering, slope-forming red shales and silt stones are called the Organ Rock Shale in Canyonlands and the Hermit Shale in Grand Canyon.

Pedestals of Organ Rock Shale (below) capped by White Rim Sandstone guard the margins of White Rim below Island in the Sky plateau. Large rocks in the foreground are collapsed pedestals, indicating the fate of the others.

WHITE RIM

As if that weren't enough, white sands again found their way into Canyonlands Country near the middle of Permian time. The deposits, known as the White Rim Sandstone, look much like the Cedar Mesa Sandstone, but the two formations are separated by the Organ Rock Shale. The White Rim Sandstone, named for the prominent topographic bench in northern Canyonlands, is thickest to the northwest, from the Green River to the San Rafael Swell, thinning to a pinchout edge along the approximate course of the Colorado River. The White Rim is present west of Meander and Cataract canyons, but is absent across the river to the southeast. It pinches out directly below Dead Horse Point, only to reappear in the north end of Castle Valley east of Moab.

The same generic problems apply to the interpretation of the White Rim Sandstone as to the Cedar Mesa. It is highly cross bedded, therefore to the "believers" the sand is windblown. However, it is definitely waterlaid, at least in part, according to us agnostics. To make a long story short, the sands were definitely windblown to the east and definitely marine toward the west. In between, the formation is a mixed breed.

There have been numerous exploratory wells drilled west of the Henry Mountains and on toward Grand Canyon, showing that the White Rim Sandstone changes into the Toroweap Formation of Grand Canyon - a marine formation. The Toroweap consists of fossiliferous limestone and dolomite that grades eastward into the white sandstone. So there is a definite marine counterpart to the White Rim Sandstone, although it occurs several miles west of Canyonlands.

Exposures of the White Rim Sandstone extend uninterrupted from beneath Dead Horse Point along the White Rim to the Green River, and southwest along the Canyonlands National Park/Glen Canyon National Recreation Area boundary to near Hite. The most interesting area of these extensive outcrops is in Elaterite Basin about 10 miles west of the confluence in Glen Canyon National Recreation Area. There, an elongate, barlike mound of White Rim Sandstone is saturated with tar and hosts numerous tar seeps. After all, the word "elaterite" means "a brownish, elastic, rubberlike, naturally occurring asphalt." In other words, it is dried-up oil.

The White Rim Sandstone is about 250 feet thick in the middle of Elaterite Basin and pinches out eastward in less than two miles, giving the formation an appearance much like an offshore sand bar. Several small canyons provide cross sections into the "bar," showing that the lower part is cross bedded, with an upper veneer of highly rippled, flat-bedded sandstone. The sand bar is more than ten miles long and oriented in a north-south configuration. Now then, how was this formed?

The most plausible explanation is that the feature is indeed an offshore sand bar, preserved in the rocks in all its glory. Keep in mind that no one knows whether or not an offshore bar should contain large-scale cross bedding, but it certainly seems likely that it should. Studies done in experimental tanks show that a man-made offshore bar looks identical to this one. However, the eolian addicts say that the "bar" is made of windblown dunes, but even they must admit that the upper veneer of symmetrical ripples was waterlaid. So they would have a windblown dune field, somebody pulls

Range Canyon between Teapot Rock and Elaterite Basin, Land of Standing Rocks in the right distance. White cliffs along the north wall of the canyon are in the White Rim Sandstone, which here pinches out from a thickness of 250 feet in the foreground to zero in the middle distance. Bench is capped by the Moss Back Member of the Chinle Formation.

the chain, and the sea comes in to reform the dunes overnight? Wow! Either way, the ocean made it into Canyonlands Country in White Rim time, even if only for a day or two.

Why the tar seeps? When oil is mixed with water the oil rises and floats on the water. All porous rocks contain water below the water table. When oil gets into a porous, wet layer of rock, it wants to rise to the top of the bed; and if the bed is inclined to the horizontal, the oil will migrate up the dip until it is trapped. In this case, the White Rim Sandstone is porous, and dips toward the west away from the Monument Upwarp. The oil somehow got into the sandstone, probably from the marine Toroweap Formation to the west, and migrated up-dip to Elaterite Basin. There it was trapped by the pinchout of the porous White Rim Sandstone, and had to just sit there and stack up in the Elaterite "bar." Now, the "stratigraphic trap" is being exhumed by erosion, presenting a "textbook" example of one kind of oil field.

At the northernmost exposure of the White Rim Sandstone at Mile 36 along the Green River, the upper part of the formation has all the earmarks of being a windblown dune sand. However, within a mile downstream, a prominent horizontal bedding plane rises to river level and the character of the cross bedding below changes dramatically. The very high angle and large size of the cross bedding in the upper formation changes to smaller-scale, low-angle, and more arcuate cross beds. It is believed that this sequence represents a barrier bar in the White Rim - an offshore sand bar that grew up to and above sea level with a final windblown dune deposit at the top (a "fossil" Padre Island). Thickness maps indicate that these exposures border a large, thick, bar-shaped deposit that trends underground to the north.

At Bonito Bend (Mile 33.5) the lower White Rim Sandstone has changed again to horizontal, crinkly beds of sandstone. This flat-bedded unit extends from the Green River eastward to where the formation pinches out under Dead Horse Point. The peculiar crinkly ripple marks, called "adhesion ripples," may indicate a sandy tidal flat environment of deposition (at least it was a wet surface for a long time) for the lower unit. The upper part of the formation is probably mostly windblown in this area. Certainly it is eolian in the exposures at the foot of Shafer Trail near the eastern pinchout edge.

The White Rim Sandstone is missing on the Moab and Salt Valley salt structures, probably because they were high and still growing during White Rim time. Drilling has shown that the sandstone is present and thick in the subsurface of the synclines between the salt structures. It is again seen as a windblown sandstone that pinches out against the northwestern nose of the Castle Valley salt structure.

If you believe all of that, our interpretation fits a plausible regional scenario: The formation is of mostly marine origin west of the Henry Mountains, where it is known as the Toroweap. A sandy marine shelf existed to the east, including the offshore bar in Elaterite Basin and the barrier bar at the Green River. And to the east, a sandy tidal flat extended as far as Dead Horse Point. Beyond that, windblown sands dominated the coastline. As the sea receded toward the west, the windblown sands followed the shoreline, covering the tidal flats with dunes.

One more advance of the sea in Permian time produced limestones of the Kaibab Formation, but they don't extend as far east as Canyonlands

Elaterite Basin. The low, white hill in the lower left of the photo is a preserved offshore sand bar in the White Rim Sandstone of Permian age. The White Rim is here saturated with tar. Beds in the lower cliff above the vehicle dip to the right, having been deposited over the sand bar and then eroded before the next higher beds of the Moenkopi Formation of Triassic age were deposited, forming an angular unconformity.

Country. The formation is present in exposures on the San Rafael Swell, the Circle Cliffs uplift, and in Grand Canyon.

UPPER PERMIAN?

Strata of Late Permian age are not known to exist on the Colorado Plateau or in Canyonlands Country. However, an enigmatic section of red beds, the Hoskinnini Member (pronounced hoss-kin-**ninney**), drives some stratigraphers half crazy. Some call it Permian in age, others think it is Triassic. Details of its relationships to other formations have not been satisfactorily worked out, but it seems that more than one rock unit has been called Hoskinnini.

"Chocolate Drops"
The Maze

CHAPTER EIGHT
MESOZOIC TIME

There was quite a gap in time between the end of Permian deposition and the beginning of Triassic time in Canyonlands Country, even though we can't pinpoint the surface in the layered rocks. At least 40 million years of Earth history are missing here. If we pretend that the Hoskinnini Member is of Permian age, the Moenkopi Formation marks the beginning of known Early Triassic sedimentation.

MOENKOPI FORMATION

Chocolate brown, thinly layered rocks above the White Rim Sandstone, or above the purplish-red Cutler cliffs to the east, represent the Moenkopi Formation (again, check the time charts inside either book cover for mental organization). The clay-rich muds were deposited on intertidal mud flats in Early Triassic time, some 240 to 245 million years ago last Thursday. It is astonishing that the coastal mud flats extended eastward a couple of hundred miles from the open sea that lay in western Utah at the time. That is one whale of a big mud flat by today's standards.

By all indications, the Moenkopi tidal flats were in every respect similar to modern-day mud flats except for the enormous geographic area they covered. Mud cracks, small water-formed ripples, raindrop impressions, and worm burrows are ubiquitous on bedding planes throughout the Moenkopi. These features are so well preserved in the laminated shales they appear to be still wet and gooey, but they are not. The "sedimentary structures" are most easily seen on slabs in stone walkways and walls of buildings in the parks and towns in the region. They originated, however, in outcrops of the Moenkopi Formation, where they can be studied in situ by the more adventurous.

The Moenkopi Formation generally thins eastward across Canyonlands, and is not present very far east of the Colorado border. The salt anticlines were again growing during deposition of the Moenkopi. The brown muds thicken in synclines between the structures, and thin or pinch out against the old high features.

There is an interesting and unique occurrence of salt within the Moenkopi in the Courthouse Wash syncline in and near Arches National Park. The syncline lies between the Moab and Salt Valley anticlines, and results from withdrawal of Paradox salt from beneath the syncline, and flowage into the anticlines. The Moenkopi salt was encountered quite unexpectedly during the drilling of a well by Shell Oil Company. The salt may have flowed like "glaciers" at the surface from one or both of the structures in

Elaterite Butte from Elaterite Basin. The flat surface in the foreground is the top of the White Rim Sandstone. Lower slope, below the bench-forming ledge of Moss Back Sandstone, is the Moenkopi Formation, and the upper slopes are in the Chinle Formation. The butte consists of the cliff-forming Wingate Sandstone, capped by a thin remnant of Kayenta Formation.

Early Triassic time, as it does in the Zagros folds of Iran today, or it may have been recycled into a salt lake in the structural valley, to be reprecipitated in Moenkopi time.

The sea once again retreated toward the west after deposition of the Moenkopi Formation, leaving the mud flats high and dry to bake in the hot sun for perhaps 10 million years. By early in Late Triassic time, streams flowing from the south again coursed across Canyonlands Country, carving shallow gullies into the Moenkopi flats. Thus, the top of the Moenkopi Formation is marked by a channeled erosional surface, or "erosional unconformity."

CHINLE FORMATION

Stream gravels and sands filled the erosional channels in the top of the Moenkopi Formation, forming discontinuous, lens-shaped bodies of conglomeratic sandstone at the base of the overlying Chinle Formation. These sandstones are called the Moss Back Member or the Shinarump Member farther south. (Although very few geologists realize it, the word "Shinarump" is pronounced shin-**air**-rump, a combination of the Paiute Indian word "shinar" meaning wolf, and we all know what a rump is. It was named for the "wolf rump" cliffs near Kanab, Utah.) The streams deposited considerable amounts of plant debris washed down from the uplands. The black, carbonized fossil remains of leaves, twigs, and even logs of trees may be found in any old uranium prospecting pit or bulldozer scar that invariably mark these beds.

Upper members of the Chinle Formation form the multicolored slopes above the Moss Back/Shinarump benches and beneath the massive, cliff-forming Wingate Sandstone. There are several member names in the upper Chinle: Petrified Forest, Owl Rock, and Church Rock to name a few, but these are difficult to distinguish in Canyonlands Country, even for the few geologists who care. Colors vary from grayish-green to pale purple, red and brown. Rock types alternate between shale, siltstone, and sandstone, and even a few limestone beds appear midway up the slopes. All were deposited by streams that came from the south, and in lakes. Petrified wood is common, some in the form of long, silicified, and usually black logs in Canyonlands. A cliff-forming sandstone in the upper part of the formation, informally called the "Black Ledge," is commonly heavily stained with desert varnish. The Chinle varies from 400 to 600 feet thick in Canyonlands Country.

The lower, slope-forming part of the formation, the Petrified Forest Member, contains large amounts of bentonitic clay derived from the weathering of volcanic ash deposits. Bentonite has a high shrink-swell characteristic, giving a frothy appearance to weathered surfaces. Many of the back roads in the region were built on this member, partly because it was easy to bulldoze the shale, and because it was immediately above the ore-bearing Moss Back Member. But beware! The bentonite content makes unpaved roads extremely slippery when wet. It pays to avoid these roads when it is raining or if rain is imminent! The back-country traveler should carry a geologic map to know where roads cross outcrops of the Chinle Formation.

Fossils may occasionally be found in the Chinle Formation. In addition to fossil plants, freshwater clams and snails have been reported, and fragmentary remains of small amphibians and reptiles occur rarely.

CHINLE URANIUM

In times when uranium mining is profitable, the Moss Back and Shinarump members are prime targets for exploration, thus the ubiquitous bulldozer scars at the lowermost fluvial layer. Sometime after the basal stream sands were deposited, groundwater leached uranium minerals from an unknown source and redistributed them by trickling through any porous rock it could find. Source of the uranium may have been the volcanic ash deposits in the immediately overlying Petrified Forest Member. The uranium remained in solution as long as the chemistry of the groundwater was acidic, as most is. However, when the fluids encountered nonacidic, or basic (alkaline) conditions, uranium oxide minerals were precipitated. Fossil plant debris in the Moss Back, Shinarump, or both, created just such reducing microenvironments within the rock, and uranium minerals precipitated in and around the plant fossils. Thus, the more fossil plants present in the rocks, the richer the uranium deposits. So fossil plants localized uranium wealth: paleontology strikes again!

Uranium prospecting ran rampant in Canyonlands Country in the late 1940s and early 1950s, as a rich market developed for the magical mineral. Many were weekend "experts," out to make a fast dollar, but others took the challenge seriously; pistol-packing prospectors crowded the streets of Moab. One Charles A. Steen fell into the latter category, catching uranium fever after a checkered early career as a free-thinking petroleum geologist from Texas. Charlie, as he was widely known, spent several years literally crawling around the ledges and cliffs of the Colorado Plateau in his dogged search for the canary-yellow wealth. He and his family were destitute and often near starvation, as Charlie persisted. Then in July 1952, while drilling a core hole on his Mi Vida claim on the flank of the Lisbon Valley salt structure, he unexpectedly encountered a rich deposit of uraninite, a then rare, black uranium ore. He formed the Utex Mining Company, and in sequence bought four airplanes, a Lincoln Continental, and built the legendary big house (Mi Vida) high above Moab, where, mostly in his words, he could "wet" on the town anytime he wanted. Charlie then sold his uranium properties to Atlas Corporation for a reported 25 million dollars. He then made and lost a few fortunes, but the big house on the hill remains as a monument to the old uranium boom days and the colorful man who started it all - Charlie Steen.

GLEN CANYON GROUP

Great Sahara-like deserts dominated Canyonlands Country during the Jurassic Period from 145 to 210 million years ago. Windblown sand dunes periodically drifted lazily across the Chinle plains, depositing sand several hundred feet thick. The first of these episodes is preserved as the Wingate Sandstone that forms the magnificent massive, vertical cliffs that effectively guard inner Canyonlands from invaders; few natural entryways exist. The Wingate cliffs also host some of the finest tapestries of gorgeous velvety, blue-black desert varnish to be found anywhere. This is the lower formation of the Glen Canyon Group, now believed to be entirely of Jurassic age after years of controversy and vacillation of the time boundary.

Although the cross bedding in the Wingate Sandstone generally indicates a windblown origin, there is evidence that in places streams were adjusting the sand deposits, and flat bedding suggests that a few seasonal lakes were scattered about the dune field. As in late Paleozoic time, the winds were blowing out of the northwest. But where did all that sand come from? That is one of the great geologic mysteries.

The Wingate Sandstone and the overlying stream-deposited Kayenta Formation cap most of the spectacular buttes, mesas and plateaus in and around Canyonlands Country. Each of the many viewpoints is atop frightening cliffs of the two formations, providing unexcelled views into the heart of Canyonlands National Park. One usually thinks of the Wingate Sandstone as the culprit forming the extensive bastions, as it is the most prominent component of the cliffs. In reality it is the Kayenta that is so resistant to the intense weathering processes of the desert environment. The formidable cliffs range in height from 300 to 500 feet, so watch that first step!

The Kayenta Formation marks a brief interval of wetter climates, as streams crisscrossed the desert, bringing more sand into play, or at least redistributing Wingate sands. As in Permian time, the streams flowed westward, probably heading in the highlands of western Colorado. Dinosaurs roamed the region as evidenced by numerous tracks and trails left by the comical reptiles. As the Kayenta sands generally thicken in the synclines adjacent to the large salt structures, it would seem that Paradox salt was again, or still, on the move.

And then it was back to the desert - and more sand! Light-colored knobs, knolls, and rounded cliffs of Navajo Sandstone stand back from the cliffs on Kayenta-capped plateaus, displaying large scale and steeply dipping cross beds of obviously windblown origin. In fact, exposures of the formation in Arches National Park have been incorrectly designated as "petrified sand dunes." The knobby terrain represents sand dunes preserved in the rocks, but there is no petrifaction involved. They have also been called "fossil dunes," but again, the term "fossil" implies the remains or traces of once living organisms.

Extensive broad plateau tops surrounding the Canyonlands are invariably capped by the Kayenta Formation, with erosional remnants of Navajo Sandstone scattered about. Examples include the approach to Dead Horse Point around Big Flat, the surface of the Island in the Sky District, the approaches to the Needles Overlook, Canyonlands Overlook and Anticline Overlook, and the high country west of the Orange Cliffs. The keys to recognizing the Navajo Sandstone are the beautifully developed and well displayed, highly complex cross bedding and its color.

The Navajo desert in those days was one of the most geographically widespread sand deposits found anywhere in the geologic record. It spread from Colorado westward into Nevada, southward into southern Arizona, and northward into Wyoming. It is known as the Aztec Sandstone in Nevada and southern Arizona, the Nugget Sandstone in Wyoming and northern Utah, and the "Nuggajo" in intermediate areas.

SAN RAFAEL GROUP

A sequence of colorful strata exposed on the San Rafael Swell west of Canyonlands Country has been lumped into the San Rafael Group of later Jurassic age. Not all formations of this group extend into Canyonlands, and those that do, the Carmel Formation and the Entrada Sandstone, have somewhat different appearances. As with the Permian formations, these change rapidly from marine deposits, eastward to sediments of continental origin.

The lower and older of these rock units is the Carmel Formation. It is largely a marine limestone to the west, as around Zion National Park, becoming gypsum-rich red beds on the San Rafael Swell. However, in Canyonlands Country the Carmel consists of reddish brown siltstones and mudstones that were deposited under marginal marine conditions. Exposures of the Carmel, especially in eastern Canyonlands and Arches national parks, are typified by highly contorted and slumped bedding above the Navajo Sandstone. Perhaps the salt anticlines were rising during deposition and shortly thereafter, and the wet muds slumped down the flanks of the growing high structures?

Unfortunately, in my opinion, the name of this picturesque formation was recently changed to the Dewey Bridge Member of the Entrada Sandstone in the vicinity of Arches National Park. Because the outcrop pattern of the red beds beneath vertical cliffs of Entrada Sandstone is very narrow when seen from above, the name was changed as a mapping convenience. The tidal flat deposits are only loosely related to the overlying cliff-forming, windblown sandstone, and really should not be made a member of an unrelated rock unit. However, it is officially known to the U.S. Geological Survey as the Dewey Bridge Member of the Entrada Sandstone in Arches National Park, and the Carmel Formation elsewhere.

Above the Carmel/Dewey Bridge red beds is a spectacular red sandstone that forms massive cliffs and hosts the many natural arches in and around Arches National Park. Now called the Slick Rock Member of the Entrada Sandstone, it used to be simply the Entrada Sandstone. The Slick Rock Member changes to a water-laid siltstone to the west in the Valley of the Goblins, and to the south as at Baby Rocks near Kayenta, Arizona. In Canyonlands Country, it is another windblown dune deposit.

A sharp, prominent bedding plane separates the Slick Rock Member from the overlying, but similar, Moab Tongue of the Entrada Sandstone in the Arches-Moab area. The Moab Tongue is thought to be a windblown equivalent of the marine Curtis Formation of the San Rafael Swell area. The Curtis is a light greenish colored sandstone that contains glauconite, a strictly marine, green, glassy iron-silicate mineral, and it contains marine fossils. The Moab Tongue is considered to be a coastal dune facies of the marine sandstone, and may be partially equivalent to the thinly bedded, chocolate brown siltstones of the Summerville Formation of the San Rafael Swell region.

MORRISON FORMATION

The youngest strata of Jurassic age in the Colorado Plateau and Southern Rocky Mountains provinces are in the Morrison Formation. It consists of three members in Canyonlands Country: the lower Tidwell, middle Salt Wash, and upper Brushy Basin. The Salt Wash and Brushy Basin members occur throughout the Colorado Plateau.

A lower red bed unit, previously thought to be the eastern extension of the Summerville Formation, is now called the Tidwell Member of the Morrison Formation. It occurs only in eastern Canyonlands Country, and is recognized as the brown siltstones and mudstones directly above the Entrada Sandstone.

Interbedded stream-channel sandstones and floodplain shales comprise the Salt Wash Member in the lower Morrison Formation. The lenticular sandstones form ledges and benches of light gray or greenish-gray color in the synclines and collapse structures within the salt valleys, being stripped from structurally high features by recent erosion. The Salt Wash Member hosts numerous uranium mines and prospects and, like the Moss Back Member, is marked by bulldozer scars, pits, and mines. The presence of fossil plant debris has localized most of the uranium deposits as in the Moss Back mineralization.

Fossil dinosaurs from the Salt Wash Member are famous worldwide. Perhaps the best known quarry is in Dinosaur National Monument near Vernal in northeastern Utah, where the huge bones and partial skeletons of the giant reptiles are being displayed as they occur in the rock. Many fossils have originated in Colorado National Monument near Grand Junction, Colorado, however, some of the best specimens have come from the little-known Cleveland-Lloyd Quarry, near Cleveland, south of Price, Utah. Wm. Lee Stokes, Professor Emeritus of the University of Utah, has supervised excavation of the fossils for many years, and has supplied rare, nearly complete skeletons and plaster replicas to museums and universities throughout the world. The working quarry is open for public visitation.

Upper slopes of the Morrison Formation are on the multicolored mudstones of the Brushy Basin Shale Member. Delicate shades of grays, greens, pinks, reds, and browns alternate promiscuously within the member. It is an obvious bright apple-green color near the junction of U.S. Highway 191 and the Dead Horse Point-Canyonlands road north of Moab. The color is probably due to the presence of iron oxides in their reduced, not oxidized, form. Recent dating (by the newly developed fission track dating method) indicates that at least the upper part of the Brushy Basin Member is of Early Cretaceous age, rather than the time-honored Late Jurassic age of older studies. If this is correct, the sequence of events remains unchanged, but the timing is altered significantly.

ROCKS OF CRETACEOUS AGE

Stream-deposited conglomeratic sandstones atop the Morrison Formation form topographic caprocks north of Canyonlands Country. The thin unit is known as the Cedar Mountain Formation west of the Colorado River, and the Burro Canyon Formation to the east. It is of Early Cretaceous age.

The Dakota Sandstone overlies the Cedar Mountain Formation north of Canyonlands Country, where it is only about 20 feet thick. It was deposited on coastal plains and advancing beaches as a great inland sea invaded the region in Late Cretaceous time. The sandstones contain fossil plant debris and a few coal beds in some areas, but is topped by marine sandstones. As with all post-Entrada formations, exposures are limited to structurally low regions, such as synclines, collapse structures, and the area north of Canyonlands and Arches national parks.

The overlying black Mancos Shale forms badlands topography along Interstate Highway 70 north of Moab. Exposures extend from near the Moab-Canyonlands Airport to high in the Book Cliffs north of I-70. Local outcrops also occur in collapse structures in the south ends of both the Moab and Salt Valley salt structures. The Mancos Shale was deposited in a relatively deep sea that extended northward from the Gulf of Mexico to Alaska, blanketing the entire length of the Rocky Mountains and eastern Colorado Plateau with dark-colored, fossiliferous marine mud. The formation is about 3,500 feet thick in the Book Cliffs area north of Canyonlands, and reaches more than 6,000 feet in thickness toward the northwest near Price. The dark gray to black color of the shale is due to a high organic content that oxidizes at the surface to a sickly yellowish color. Bentonitic clays, like those in the Chinle Formation, make the Mancos Shale nasty and unstable for road and airport construction, although providing easy routing. Paved surfaces tend to sink out of sight, and unpaved roads entrap vehicles as if in wet glue after a rain. Locals call it gumbo. Dirt roads built on the Mancos Shale are treacherous when wet!

Gray sandstone cliffs high in the Book Cliffs are beach and nearshore marine sands of the Mesaverde Group. They record advances and retreats of the Late Cretaceous shorelines as relative sea level was again in a state of flux. These beds have been eroded from most of Canyonlands Country.

Very thick stream and lake deposits of Tertiary age occur in the Uinta Basin north of the Book Cliffs. Although these rocks may have originally blanketed all or parts of Canyonlands Country, they have been removed by recent erosion and will not be discussed here.

CHAPTER NINE
OROGENOUS ZONES

Big things happened to the Colorado Plateau Province, and Canyonlands Country in particular, around the end of Cretaceous and the beginning of Tertiary times. Some would say that it was the most significant event in geologic history. A great wave of compressional Earth forces rolled eastward through the crustal rocks of western North America, buckling and smashing the rocks against the more rigid core of the continent. We call this event the Laramide Orogeny; it happened some 60 to 70 million years ago.

Geologic structures already in place since Precambrian time were generally elevated, enlarged, and pushed over toward the east to form the great uplifts and monoclines we see at the surface today. The Monument Upwarp and San Rafael Swell are the most noticeable of the huge folds in and around Canyonlands Country, but the Uncompahgre Uplift was elevated again along the old faults. In general, the uplifts became higher, the basins relatively lower, and the smaller structures were enhanced. The crust was shortened by several miles from west to east in the process. The old vise handle was given another turn, but this time the Colorado Plateau megablock got it from the west instead of the north or south.

Some geologists blame all of the structural blemishes of the Colorado Plateau on the Laramide Orogeny. That's easy to do, because these surface structures are the most obvious. As we have already seen, however, all of the "Laramide structures" were present in some form since the Precambrian, some billion and a half years before Laramide time. The old structures were merely exaggerated, elevated, and rotated eastward by the later compressional event. "Old structures never die..." There's nothing new under the desert sun!

But what could cause such a mess? According to the new global tectonics scheme, the North American plate (continent) began moving westward relative to the Pacific oceanic plate in Jurassic time. The continental plate, being lighter, began to override the denser oceanic plate, causing the Pacific plate to dive under, or be subducted. Great eastward-directed compression buckled the continental crust, i.e. friction is a drag.

Compressional structures began to form in eastern California and western Nevada in Jurassic time, becoming younger as the process migrated eastward. By latest Cretaceous to earliest Tertiary time, the compressional wave reached the Colorado Plateau and reactivated existent geologic structures; the Laramide Orogeny did its job.

EPEIROGENY

It all began in Early Tertiary time, when the Colorado Plateau was bodily tilted toward the north. Erosion began stripping the higher country to the south, and drainage proceeded to carry sediments northward into Lake Uinta north of Canyonlands Country. The erosional stripping of the Plateau was underway; the resulting sediments were dumped into the huge lake for quick and dirty disposal.

By Middle Tertiary time, the entire midsection of the continent, including the Rocky Mountain and Colorado Plateau provinces, was being elevated en masse to several thousand feet. This kind of mass elevation of a continent is called epeirogeny (pronounced eh-pie-**roj**-any). Now erosion began in earnest, but the direction of drainage essentially reversed itself. The master rivers began to flow south and west toward the Pacific Ocean, slowly but surely establishing their present-day drainage patterns. Vast quantities of sediment were removed from the Colorado Plateau, this time heading for the Gulf of California.

Meanwhile, compressional forces acting on western North America relaxed, and the Basin and Range Province to the west and south of the Colorado Plateau broke apart. Up-faulted mountain ranges ("horsts") and down-faulted valleys ("grabens") resulted from the new tensional regime, forming the typical landscapes of today. The land was actually being stretched, rather than squeezed, by relaxation of Laramide forces.

On the Colorado Plateau and Canyonlands Country, the relaxation of compression allowed the deep-seated basement faults to open up a bit, and igneous intrusions began working their magma (molten rock) up into the crustal rocks about 25 million years ago. In places where magma reached the surface, volcanoes erupted. Elsewhere, the magma cooled much more slowly underground, and intrusive igneous rocks resulted. These features became the laccolithic (or cactolithic) bodies of igneous rock that would form the mountain masses (La Sal, Henry, Abajo, Ute, and La Plata ranges, and Navajo Mountain) when later exhumed by erosion.

Finally the stage was set to carve the magnificent scenery of Canyonlands Country!

CHAPTER 10
CANYON CUTTING

Now that we have all of the rocks in place, we can start tearing them down.

It rains! A trickle of water makes its way down any slope, trying to accommodate gravity. It picks up a grain or two of sand, a little mud, and carries it along toward the ocean. The next time it rains, there is the little rivulet formed by the former runoff, and this rain water follows the same course, as it is the easiest route. Before long there is a small gully, then a gulch, a ravine, and finally a canyon, as each rain enlarges the original water course.

Running waters carry bits of rock, mud, sand, and gravel that they pick up along the way. The sediments in turn aid the canyon-cutting process by acting as grinding tools on the stream bottom. The action is like a buzz saw travelling down river on a conveyor belt. And the removal of the stream sediments lowers the land surface and deepens the canyons. The process is endless and relentless! But there has been plenty of time!

Rivers are more effective at erosion when they are high above their ultimate destination - the ocean, and when the slope, or gradient of the river's course is steep. In this case, running water moves faster, carries more sediment, and consequently cuts its canyon faster. The lowest level to which a river can cut its course and still flow is called its "base level of erosion." As the river bottom approaches this base level, it becomes sluggish, carries only the finest sediments, and downcutting comes to an end.

The rivers of the Colorado Plateau are high in elevation above sea level, their waters can be fast, they can carry copious amounts of sediments, even move boulders, and the canyons are deep and getting deeper. Erosion here is working at its best, and the canyons are at their best.

Of course, the reservoirs behind the man-made dams don't help. Each one becomes a temporary base level of erosion, stopping downcutting and infilling the original canyons with sediments. Not only that, the dams control the rivers' flow in times of flood, thus the erosional process below each dam is greatly hampered. Dam-building is another of man's insidious attempts to fool with natural processes that will come back to haunt us in the end. Just wait until one of these monstrous dams fails! And Glen Canyon Dam nearly did just that in 1983, when an unexpectedly high runoff nearly took it out.

But why is a canyon formed in any particular location? One answer is obvious: the unique route of a canyon was the easiest course for a stream to follow when it first began to flow. As one would guess, geologic structure should play a large part in producing the easiest course. Rivers should flow around uplifts and into basins, or at least flow down the dips of the rocks. But

Lower Cataract Canyon near Bowdie Canyon, now inundated by Lake Powell. The lower cliffs are rocks of the Honaker Trail Formation and the middle ledgy slopes are in the Halgaito Shale, a lateral equivalent of the Elephant Canyon Formation. The high, light-colored cliffs are in the Cedar Mesa Sandstone.

the Green and Colorado rivers didn't know that! They took on every challenge in sight and conquered uplifts and basins indiscriminately.

We have seen that the Colorado Plateau was tilted bodily northward in early Tertiary time, only to be vertically uplifted in the middle Tertiary. As such, the entire Colorado-Green drainage system flows southward against the tilted structural grain, and crosses nearly every uplift on the Colorado Plateau. The rivers don't flow around the structures, as one would expect, but cross them as if they didn't exist. The Colorado River crosses the Uncompahgre Uplift at right angles, then flows across the nose and along the flank of the Monument Upwarp. The San Juan River crosses the girth of the Monument Upwarp at right angles. Then the combined rivers cross the Kaibab Uplift at right angles in Grand Canyon. Meanwhile the smaller rivers dissect the smaller structures. For example, the Dolores River crosses the Gypsum and Paradox Valley salt anticlines at right angles before joining the Colorado River near the Uncompahgre Uplift. Something is wrong here!

Attempts to clarify the enigma go back to Major John Wesley Powell, who first noticed these and other incongruities during his exploratory trips of 1869 and 1871. He believed that the structures were growing across the already established river courses, that downcutting by the rivers was able to keep pace with the growth of the structures, and thus they maintained their original courses. Consequently, he called these canyons "antecedent." We now know, however, that the major uplifts and basins formed long before the rivers took their present courses, although mass uplift of the Colorado Plateau may still be in progress.

Powell also realized that the courses of the rivers may have been established on a relatively flat surface on younger rocks. These younger strata may have buried the existing structures, and later were eroded. Shortly after deposition of the younger layers, regional uplift occurred, streams began downcutting into the soft strata and soon became trapped in their own canyons. When they had cut down to hard rocks on the older structures, they couldn't escape their own canyons, and incised themselves into the hard rocks. Powell called such canyons "superimposed," or let-down from above.

Charlie Hunt, in his classic studies of the Colorado River drainage system, noted that each canyon on the Colorado Plateau has a unique but related history. It is beyond the scope of this book to argue the case for each canyon. However, in Canyonlands Country, it is safe to assume that the canyons are of the superimposed type. The rivers took their approximate present courses on flatter surfaces above the Cretaceous rocks, long after the structures were formed by the Laramide Orogeny. They meandered about this low-lying plain until epeirogenic uplift began. Then with renewed vigor brought on by steepened gradients, the rivers cut down into preexisting structures, becoming hopelessly incised. Numerous incised meanders and cut-off meanders along the major rivers on the large uplifts would seem to preclude other interpretations in this region.

There is an interesting hint of structural control on the course of the Colorado River, which flows amazingly straight southwestward from about Grand Junction, Colorado to the east end of Grand Canyon, Arizona. Geophysicists with the U.S. Geological Survey realized in the early 1960s that a major break in the magnetic nature of the basement rocks occurs across this general line. They later named this apparent deep-seated fault the

Colorado Lineament. Because of nomenclatural complications, we now call it the Cataract Lineament. The only significant variation of the river from this straight course is at Mille Crag Bend near Hite, where the river turns sharply to the west for 10 miles, and then reverts to its straight southwesterly course. This is exactly where the Cataract Lineament is displaced in a right hand sense along a northwest-trending structural zone called the Four Corners Lineament. (These relationships are shown on the map, page 26.) East-west faulting is obvious along the canyon walls in this case. Otherwise there is no apparent relationship between basement structure and surface control of the river's course. Yet, for some obscure reason, the course of the river mimics the deep-seated fault or faults. This phenomenon is not uncommon elsewhere in the world.

Canyon widening is not a function of downward cutting by a river, but is caused by mass wasting of the canyon walls. In other words, rock falls, mud flows, flash floods, slumping, all cause canyon walls to recede and broaden the canyon. All of these factors are active and apparent within Cataract Canyon, but in this case rocks of the inner gorge are very hard and resistant to widening processes. The massive Cedar Mesa Sandstone holds up the canyon rims, but softer rocks above the Cedar Mesa, the Moenkopi and Chinle Formations, have been eroded back for five to 15 miles from the inner canyon to the prominent Wingate-Kayenta cliffs beyond. The sharp inner canyons are being cut into older, much more resistant limestones, where canyon-widening processes are very slow in comparison.

On a more detailed level, the reasons why a river turns exactly where it does, or why an erosional remnant forms a butte in a specific location are more difficult to explain. Sometimes these features are controlled by fracturing in the rocks, as in the Needles District of Canyonlands National Park, but often there are no recognizable specific controls. A fine geologic term explains all such occurrences with authority: "differential erosion."

SUMMARY

Canyonlands Country is underlain by a highly fractured basement of brittle metamorphic rocks, that was buried in early Paleozoic time by widespread marine sedimentary rocks. Basement faults were reactivated in Pennsylvanian time, forming a large, down-faulted trough, the Paradox basin, bordered on the northeast by a major mountain range, the Uncompahgre Uplift. The basin was filled with sediments derived from the uplift on the east, as thick salt accumulated throughout the rest of the basin. Marine limestones accumulated along the seaward shelves of the basin, including oil reservoir rocks. The basin was completely buried by red sediments from the highlands in Permian time, forcing the seas to retreat toward the open ocean to the west.

Triassic time began with widespread tidal flats of the Moenkopi Formation, later to be buried by stream and lake deposits of the Chinle Formation. Great sand deserts dominated the region in Jurassic time, leaving hundreds of feet of dune sand deposits of the Glen Canyon Group that form cliffs that effectively guard entry to inner Canyonlands. Stream deposits of the Morrison Formation supported thousands of dinosaurs near the close of the

Jurassic Period. Uranium deposits accumulated in stream sands of the lower Chinle and Morrison Formations.

Tectonic unrest, starting in the west in Late Jurassic time and later migrating into Canyonlands Country, crowded the main seaway from western Utah into the Midcontinent during the Cretaceous Period. It culminated in the Laramide Orogeny by the Early Tertiary, generally uplifting and exaggerating preexisting structures, and forming the great uplifts and monoclines of the Colorado Plateau. The entire province was tilted toward the north, then bodily uplifted several thousand feet, to initiate erosional stripping of Canyonlands Country. This is the legacy of past geologic time.

Erosion is continuing today at a feverish pitch, verifiable by anyone who has seen a July cloudburst or Cataract Canyon in flood. Geologic time marches on!

Buttes of the Cross
Green River

PART TWO
GEOLOGICAL TOURS

Druid Arch

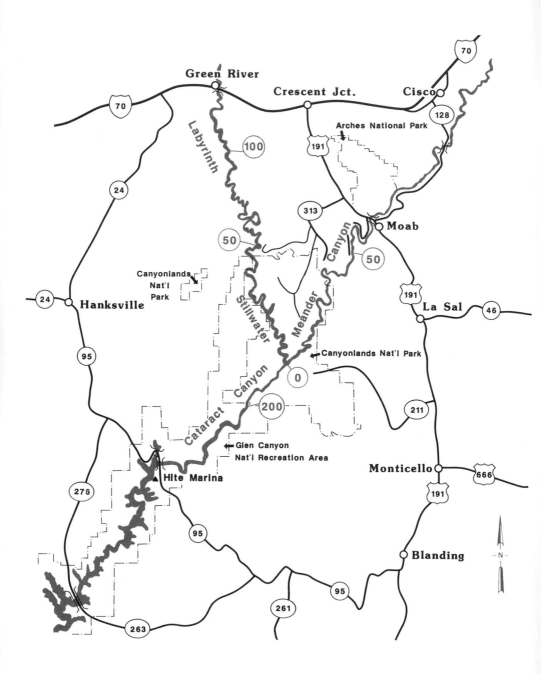

Map showing major access roads to Canyonlands Country.

Part Two is a geological tour guide that may be used separately, especially for those familiar with basic geological concepts, without the more detailed historical story told in previous chapters. This section deals with local features seen at the surface and does not delve into background material. For a better understanding of the "why's and wherefore's," the reader should refer to Part One - a short course in geology, at least as it applies to Canyonlands Country.

Let's take a closer look at the geology of Arches and Canyonlands national parks, one district at a time. We will begin each tour at Moab, Utah, for geographic convenience, and describe features of interest en route to the various areas. Perhaps the best place to begin is Arches National Park, as it is the most accessible and to many, the most fascinating. So let's begin our journey into the magnificent land of "...ten thousand strangely carved forms."

MOAB

The area around the town of Moab is geologically fascinating. Moab Valley is the surface expression of a salt-intruded anticline, one of several in the eastern Colorado Plateau country. There is a major deep-seated fault underlying the western valley wall, along which salt from the Paradox Formation has flowed upward. There is no salt present under the west rim, as shown by drilling, but there may be as much as 12,000 to 15,000 feet of salt beneath the valley floor. These relationships are shown in the diagram on the next page.

In the course of the upward movement of salt, the overlying strata have been punctured and bowed upward to form the overall anticlinal structure. Circulating groundwaters have removed some of the salt near the surface in the past few million years, and the overlying sedimentary rocks have collapsed into the solution void. Collapsed formations are especially obvious at either end of the straight, northwest-trending valley. The eastern valley wall is also a slump feature formed by collapse.

Rocks exposed in both walls of Moab, or Spanish Valley, are Navajo Sandstone at the top, seen as a light-colored, rounded-weathering cliff. The Navajo is underlain by the brown, more ledgy sandstone beds of the Kayenta Formation and the massive brown cliffs of the Wingate Sandstone, all of Jurassic age. Because of the collapsed top of the salt structure, these formations are highly fractured and badly slumped in the western cliffs. Lower slopes in the valley walls are in the Chinle Formation of Triassic age.

On either side of The Portal, where the Colorado River exits the valley toward the west, are exposures of upturned limestones of the

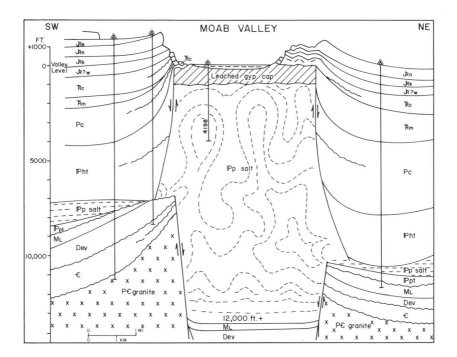

A diagrammatic sketch showing what the rocks may look like beneath the Moab salt-intruded anticline. Major deep-seated faults are known to underlie the flanks of the up-arched structure. They first moved in Precambrian time, and moved repeatedly later in Earth history as well. The vertical lines represent deep wells that help resolve the nature of the structure. Salt flowed from the flanks toward the faulted structure and then upward over a period of about 160 million years, piercing the overlying rock layers. Near-surface ground water later dissolved the topmost salt, leaving behind a cap of leached gypsum, shale, and limestone, and overlying rock layers collapsed back into the remaining void.

(Symbols indicate the ages and names of the rock formations. PC=Precambrian, C=Cambrian, Dev=Devonian, Ml=Mississippian Leadville Limestone, lPpt=Pennsylvanian Pinkerton Trail Formation, lPp=Pennsylvanian Paradox Formation, lPht=Pennsylvanian Honaker Trail Formation, Pc=Permian Cutler Formation, TRm=Triassic Moenkopi Formation, TRc=Triassic Chinle Formation, JR?w=Jurassic? Wingate Sandstone, JRk=Jurassic Kayenta Formation, JRn=Jurassic Navajo Sandstone, and JRe=Jurassic Entrada Sandstone.)

Pennsylvanian Honaker Trail Formation, directly overlain by brown siltstones of the Moenkopi Formation. The entire Cutler Formation is missing because the salt structure was high and growing during Permian time. Another angular unconformity within the Chinle Formation at the south side of The Portal registers another movement of salt during Triassic time. Numerous low, rounded gray hills along the edges of the valley floor are remnants of gypsum beds that came up with the salt and now form a residual gypsum cap on top of the salt structure. The time and rock charts inside either cover are handy references for sorting out the sequence of formations and ages.

Aerial view looking southeastward across Moab Valley to the La Sal Mountains. Moab Valley is the collapsed crest of a salt structure, formed by upward piercement of the layered rocks by flowing salt. The Colorado River crosses the valley at right angles in the near distance.

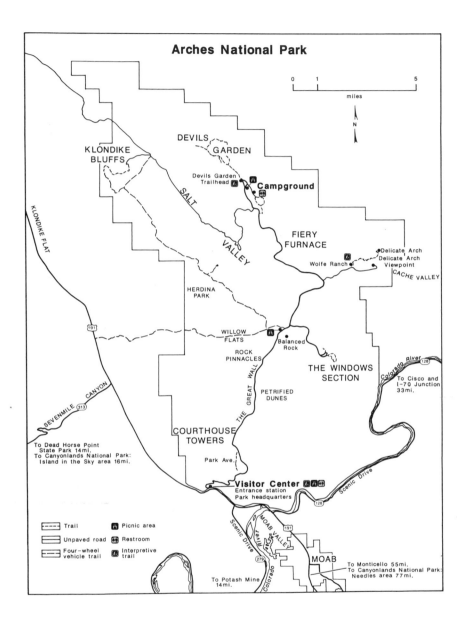

Arches National Park

0   1   5

miles

N

KLONDIKE
BLUFFS

DEVILS GARDEN

Devils Garden
Trailhead

Campground

SALT VALLEY

FIERY
FURNACE

Delicate Arch
Delicate Arch
Viewpoint
Wolfe Ranch

CACHE VALLEY

HERDINA
PARK

KLONDIKE FLAT

191

WILLOW
FLATS

Balanced
Rock

ROCK
PINNACLES

THE WINDOWS
SECTION

Colorado River

128

To Cisco and
I-70 Junction
33mi.

PETRIFIED
DUNES

THE GREAT WALL

SEVENMILE CANYON

313

To Dead Horse Point
State Park 14mi.
To Canyonlands National Park:
Island in the Sky area 16mi.

COURTHOUSE
TOWERS

Park Ave.

Visitor Center
Entrance station
Park headquarters

Scenic Drive

128

MOAB VALLEY

191

Scenic Drive

Colorado River

279

MOAB

To Monticello 55mi.
To Canyonlands National Park:
Needles area 77mi.

To Potash Mine
14mi.

Colorado River

Trail

Unpaved road

Four-wheel
vehicle trail

Picnic area

Restroom

Interpretive
trail

CHAPTER 11
ARCHES NATIONAL PARK

There is something magical about natural arches. They come in myriad shapes and sizes, yet all arouse human curiosity to the limit. How do they form? Why are they here? Why don't they fall down? How many can I photograph with the spouse and kids underneath in a day?

Arches National Park is said to have the greatest density of natural arches of any similar geographic area in the world. It certainly has the greatest variety and the most picturesque arches to be found anywhere. Yet natural arches are merely geological curiosities. Their significance lies in the story behind the unique concentration of arches at this particular place.

There are two fundamental prerequisites for the formation of natural arches. First, a mechanism must exist to cause the rocks to erode into narrow vertical slabs, or "fins." Second, the rocks involved must be strong enough to stand without support if underlying rocks erode or fall away. In Arches National Park, fins are wondrously numerous, formed by fracturing of the rock layers along the flanks of Salt Valley, and the rock is, by erosional coincidence, the Entrada Sandstone, a relatively hard rock layer. So the reason why there is such a concentration of arches here has something to do with the geology of Salt Valley.

SALT VALLEY

Only one of many in the eastern Colorado Plateau Province, Salt Valley is a salt-intruded anticline, or salt "diapir" (pronounced dy-a-**peer**). The several salt structures occur along the trend of thickest salt in the Paradox basin, a long, down-faulted basin of Middle Pennsylvanian age. Deep drilling and geophysical studies of these structures have helped explain their localization and development.

Each of the salt structures is underlain by one or more very large, deep-seated faults, and Salt Valley is no exception. The faults were active more than a billion years ago, in Precambrian time, and were rejuvenated during the Pennsylvanian Period, some 300 million years ago. They were active elements in the subsidence of the Paradox basin, where sea water was entrapped and salt precipitated over millions of years time. The vertical displacement along these faults was on the order of a mile or more, as documented by drilling. Within any one of these down-faulted blocks 4,000 to 8,000 feet of salt was originally deposited. A more detailed account of the origin of the Paradox basin may be found in Chapter Six.

Meanwhile, the ancient Uncompahgre Uplift was rising along faults to the east, and thousands of feet of sand, cobbles, and mud from the ancestral mountain range were being deposited onto the flatlands below. The great volume of sediments was deposited along and onto the bedded salt, creating a tremendous load on the plastic salt. As the mobile salt flowed westward away from the lopsided overburden, it encountered the large, rigid fault blocks and was deflected upward. Once the salt began its upward flow, nothing could stop it. Salt flowage first bulged the overlying rocks, and later actually pierced the layers, perhaps flowing out onto the surface at times. There may now be as much as 15,000 feet of salt underlying Salt Valley.

The salt structures grew from Middle Pennsylvanian through Jurassic times, a period of some 150 million years or more. The Salt Valley structure was growing upward as the Entrada Sandstone was being deposited and cemented into rock, and after the sandstone had become brittle as well. Intense bulging shattered the rock, forming a multitude of elongate fractures parallel to the wall of mobile salt. The fractures opened further when groundwater dissolved some of the near-surface salt, and overlying beds, including the Entrada Sandstone, collapsed into the void to form the valley over the wall of salt.

So Salt Valley is the collapsed top of a salt-intruded anticline. Intense fracturing of the brittle Entrada Sandstone formed the many fins, and normal weathering processes attacked and enlarged the fractures. The rock comprising the fins, the Slick Rock Member of the Entrada Sandstone, was sufficiently well cemented and strong enough to stand against the elements, with or without complete underlying support.

MAKING ARCHES

Now that we have a multitude of narrow fins of hard rock along both flanks of Salt Valley, we can begin shaping the fins into arches and other related landforms. Only a bit of undercutting is needed to allow chunks or slabs of sandstone to fall from the cliffs. If a bed of shale, ever so thin, or a layer of poorly cemented sandstone, is interlayered in the Entrada Sandstone, it will be more susceptible to weathering and will start the undercutting process.

Arch formation began, in many cases, at the contact between the lower Dewey Bridge Member and the massive Slick Rock Member. The brown, contorted siltstone of the Dewey Bridge erodes much more rapidly than the sandstone, and a notch may form readily. When this happens, the Slick Rock sandstone simply begins to flake off, and an arch is started. The sandstone then will spall from the cliffs along conchoidal fractures, much like the curved fractures made when flint or glass is flaked.

The sandstone has been covered by thousands of feet of rocks for millions of years. When bared by erosion, the release in pressure causes the rock to spall, or flake off. Flaking starts at the lower bedding plane and curves upward and outward to form a cove. When the cove breaks through the fin, an arch results.

Arches usually start small and expand in size by gravity collapse or spalling processes, enhanced by natural weathering. And yes, the process will continue until the arch eventually falls. Arches in all stages of development may be found within the park, from small, simple collapse

Delicate Arch, Arches National Park. The fragile feature is a relic left by the erosion of the host Entrada Sandstone. The steep dip of the rocks beyond the arch resulted from collapse of strata into the Salt Valley salt structure. La Sal Mountains, shrouded with their normal complement of thunderstorms, lie in the distance.

"windows" above shale beds or bedding planes, to large, graceful expanses, and finally to collapsed arches. More than 500 arches have been reported within Arches National Park.

Natural bridges, by contrast, form over incised stream courses, while arches do not. Bridges are commonly formed where an entrenched meander along a water course is undercut until stream erosion breaks through the "gooseneck." The runoff can then flow beneath the natural bridge, and a cutoff meander results. Spectacular examples may be seen in Natural Bridges National Monument, between Blanding and Hite, Utah.

BUT WHY HERE?

We have seen that several salt-intruded anticlines, in most ways identical to Salt Valley, are distributed along a trend in the Paradox basin. The flanks of other structures are equally as fractured as the flanks of Salt Valley. Why then are the arches concentrated here?

In the first place, hundreds of arches are scattered throughout the Colorado Plateau Province, but nowhere is the concentration as dense as in Arches National Park. Second, Salt Valley is the only salt structure where the Entrada Sandstone is the dominant rock exposed high on the structural flanks. The Navajo Sandstone is the flanking rock along Moab Valley; Cutler sandstones flank the Onion Creek-Fisher Valley structure; and the Wingate Sandstone caps Castle Valley, Paradox Valley, and Sinbad Valley structures. The quirks of erosional stripping of the various salt structures have made Salt Valley unique in this regard.

So what, then, is so special about the Entrada Sandstone? That is one of the great unsolved mysteries of the region. Perhaps the Entrada is more brittle due to the nature of the cement holding the sand grains together. Perhaps the nature of the beds below or above the formation is involved. Perhaps it has something to do with microclimates that control local weathering processes. If you find the answer, let us all know!

SEEING ARCHES NATIONAL PARK

EN ROUTE

Travelling north from Moab on U.S. Highway 191, the first prominent landmark is the large house (Mi Vida) sitting high on the valley wall to your right. It was built by Charlie Steen who discovered the Mi Vida uranium mine in 1952, opening the Big Indian Mining District south of Moab. Then on the left is the Suburban Gas Company that stores L.P.G. (Liquid Petroleum Gas) 2,000 feet underground in a cavern washed out of the Paradox salt.

Before crossing the Colorado River bridge, you can see on the cliffs to the right beautiful exposures of the Glen Canyon Group of Jurassic age. The upper, rounded, white cliffs are Navajo Sandstone, the ledgy, more thinly bedded, brown cliffs are the Kayenta Formation, and the lower massive cliffs are the Wingate Sandstone. The underlying Chinle Formation is exposed in roadcuts just beyond the bridge. From this viewpoint, the Glen Canyon Group rolls over toward the west in a crumbling heap of rock that has collapsed into

Moab fault at entrance to Arches National Park. The smooth dark boulder-strewn slope between the dark lines is the Moab fault zone. Rocks to the right of the fault are in the Entrada Sandstone; rocks to the left are in the Honaker Trail Formation.

the northwest end of the Moab salt structure. The angular unconformity between slightly upturned beds of the Honaker Trail Formation and the overlying, more flat-lying brown Moenkopi Formation can be seen across the valley, low in the cliffs. The unconformity represents a gap in the rock record of perhaps 50 million years. The Cutler Formation is not present here due to growth of the salt structure in Cutler (Permian) time.

The Colorado River flows directly across Moab Valley, instead of running down the valley as would any sensible river. The river established its course when several thousand feet of younger strata buried the salt structure: the river was "let down" onto the structure as erosion progressed, making the Colorado a "superimposed" river. Erosional processes are described in Chapter 10. This is the approximate site where the Old Spanish Trail crossed the Colorado River.

Across the bridge on the left is the Atlas Minerals uranium concentration mill and tailings pond, followed by the paved road to the Potash processing plant.

THE PARK

The entrance to Arches National Park is just off U.S. Highway 191, five miles north of Moab. The highway runs along the Moab fault at the entrance junction. The fault may be seen just to the north where the Slick Rock Member of the Entrada Sandstone on the east (right) has been faulted down against the Honaker Trail Formation, a vertical movement of about 2,500 feet. A terrible jumble of boulders and slabs of Navajo Sandstone forms the hillside just south of the junction, where the rocks have collapsed into the north end of the Moab salt structure.

From the Park Entrance and Visitor Center, the road winds up the folded top of the Navajo Sandstone, along a bench between the underlying Navajo and the softer Dewey Bridge Member of the Entrada Sandstone. It will follow this natural terrace all the way to Balanced Rock, a distance of about nine miles. The buff-colored, highly-cross bedded sandstone to the right of the road is the Navajo, and the reddish brown rocks to the left and above the road are the Entrada Sandstone.

Crinkly bedded mudstones and siltstones of the Dewey Bridge Member of the Entrada (formerly known as the Carmel Formation) form the base of the brown cliffs to the left. The rocks were deposited as tidal flat muds in Jurassic time marginal to a seaway that lay to the west. The contorted bedding probably formed as the soft muds slumped down along the tilted top of the Navajo Sandstone along the flank of the growing Salt Valley structure. It must have happened shortly after deposition as the mud was being buried by windblown sands of the Slick Rock Member. Delightful buttes and pinnacles of the Entrada Sandstone decorate the landscape from Park Avenue to Balanced Rock, including an anatomically correct "male organ" along the way. The term "Petrified Dunes" for the hummocky terrain to the right is technically incorrect, but conjures up the proper image of windblown dunes cemented to become the Navajo Sandstone.

A short 2.5-mile drive to the south from Balanced Rock leads to a bevy of beautiful arches, known as the Windows Section, formed in the

Entrada Sandstone.

From the Windows junction, the main road descends the west flank of the Salt Valley salt structure. Good views of the elongate, northwest-trending, collapsed salt anticline may be seen toward the north (left). A dirt road turns right from the main road near the base of the hill, leading to Delicate Arch trail and viewpoint, and Wolfe Ranch through upper Cache Valley. Rocks along the way are jumbled masses of Morrison Formation (greenish-colored shales and sandstones), the Dakota Sandstone, and dark gray masses of Mancos Shale, all of which have collapsed in a heap into the dissolved top of the salt structure.

After hiking to Delicate Arch, return to the main road and continue to the right. The road parallels huge blocks of Morrison Formation, Dakota Sandstone, and Mancos Shale that have collapsed into the valley floor past the Fiery Furnace Viewpoint. Then the road climbs into the shattered mass of Entrada Sandstone, eroded into jillions of fins in the vicinity of the Devils Garden trailhead and campground. The magnificent array of fins really should be seen from the air to be appreciated.

It is apparent that arches are easily formed from these narrow walls of rock, and indeed the best ones occur in this area. Hike the well marked trails to any or all of the local arches, but at least walk to the longest arch in the park, Landscape Arch, with an unbelievable span (opening) of 306 feet. Be sure to carry a more-than-adequate supply of drinking water on any hikes, especially in the hot summer months.

If you are inclined to camp, make sure space is available in the very pleasant Devils Garden Campground as you enter the park. If not, return to Moab and excellent accommodations by the route you came in on. It is possible to exit Arches National Park by a rough dirt road that heads north up Salt Valley to near Crescent Junction, but four-wheel-drive is recommended.

Skyline Arch

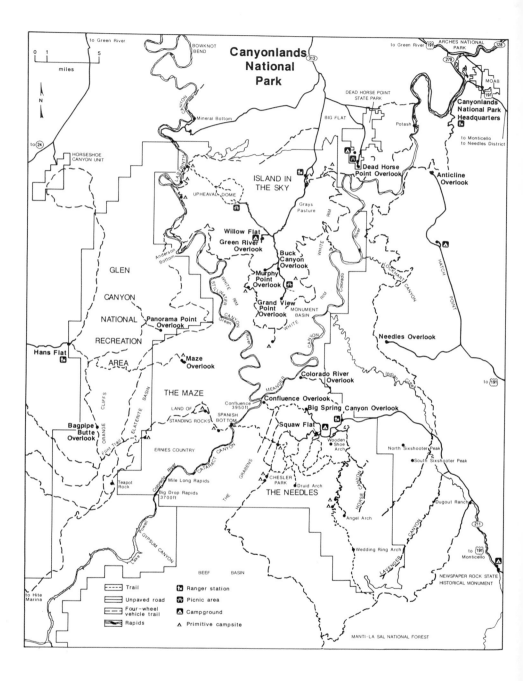

Canyonlands National Park

CHAPTER 12
ISLAND IN THE SKY

Follow the description of the route from Moab northward to the Arches National Park entrance, a distance of about five miles, and continue on northward on U.S. Highway 191.

Just beyond the entrance to Arches National Park, the highway crosses the Moab fault. The Entrada Sandstone, ahead to the right of the fault, was dragged down and faulted against the Late Pennsylvanian Honaker Trail Formation on the left. Limestones in the Honaker Trail here contain abundant fossils of crinoids, bryozoa, brachiopods, horn corals, and a few trilobites and microfossils. Stop in one of the big parking spots and look for fossils in the rocks forming the low, gray ledges along the wash.

The high cliffs on the left after the S-turn are capped by the Wingate Sandstone, underlain in turn by the Chinle and Moenkopi formations. These formations are described in Chapter Eight. The lower, bright red cliffs are the Cutler Formation, thickening from its pinchout against the Moab salt structure. The formation is here only about 400 feet thick, but it thickens along the cliff to about 800 feet at the last visible exposures ahead (north). These rocks of Permian age are discussed in Chapter Seven. The highway parallels the Moab fault to the Dead Horse Point-Canyonlands junction.

About 11 miles north of Moab turn left to Dead Horse Point and Canyonlands National Park. The road crosses railroad tracks leading to the potash mine down-river from Moab, and again crosses the Moab fault. Here green shales of the Morrison Formation are faulted against Cutler red beds at the base of the cliff. The high cliffs are Wingate Sandstone overlying Chinle and Moenkopi shales. The rocks are dipping toward the west, away from the fault, so the road gradually goes up through the rock section. First the Cutler goes underground, then the Moenkopi and Chinle disappear, followed by the Wingate. Then the road winds its way up through ledgy cliffs of the Kayenta Formation.

When the road rounds the upper sharp curve to the rim, two buttes are visible across the canyon to the right. Named the Monitor and the Merrimac for the Civil War battleships; they are in the Entrada Sandstone. The Dewey Bridge Member forms the contorted lower beds, the Slick Rock Member forms the massive cliffs, and the Moab Tongue makes the upper, lighter-colored cliffs above the prominent bedding plane. The Morrison Formation can be seen in the higher hill to the left of the buttes.

The road is in the King's Bottom syncline. The formations can be seen rising toward the west onto the Cane Creek anticline. White hills and knobs of highly cross-bedded Navajo Sandstone occur from here to the road's end.

Turn left to Dead Horse Point State Park, 13.7 miles from U.S.

Grand view into Canyonlands from Dead Horse Point. Looking toward the southwest, the Colorado River wends its way through Meander Canyon. Red rocks from the river up through the first bench of light-colored limestone are in the Elephant Canyon Formation. Reddish-brown cliffs above the low bench are in the Cutler Formation. A red slope-forming layer beneath the obvious White Rim bench, midway up the section, is the Organ Rock Shale. The White Rim Sandstone forms the White Rim, a prominent bench that extends far into the distance. Dark brown ledgy slopes immediately above the broad White Rim bench are in the Moenkopi Formation, capped by the lighter-colored slopes on the Chinle Formation. Finally, the massive, near-vertical brown cliffs at the top are the nearly impenetrable bastions of the Wingate Sandstone, capped by the more ledgy Kayenta Formation. Dead Horse Point is capped by the Kayenta Formation.

Highway 191. The junction is at the edge of a broad, grassy plain on the Big Flat anticline. This was the site of the Big Flat oil field, discovered in 1957 by Pure Oil Company (now Chevron), that produced 83,000 barrels of oil from limestones of Mississippian age at a depth of about 7,700 feet before mechanical problems caused the field to be abandoned. The anticline is so broad and gentle at the surface that it is not noticeable.

Stop and browse at the Visitor Center as you pay the entrance fee to Dead Horse Point State Park.

DEAD HORSE POINT

The view into Canyonlands Country from Dead Horse Point is nothing short of spectacular! Looking south, the Abajo Mountains rise above the far horizon, and the broad curved skyline to the right is formed by the crest of the Monument Upwarp. In the middle distance, the ragged terrain of the Needles District in Canyonlands National Park is visible. Directly below the point, the Colorado River makes its way through the heart of Shafer Dome, a northeast-trending anticline that is difficult to delineate from this steep angle of view. The inner canyon is cut into the Elephant Canyon Formation of Permian age. Grassy slopes and flats surrounding the inner gorge are on the highest limestone of the formation, here informally called the "Shafer lime." These rocks are described in Chapter Seven. You are standing on the Kayenta Formation, and the massive vertical cliff below is Wingate Sandstone.

Now stroll along the path to your right, but hang onto the kids! The view in that direction shows the sequence of the stratified rocks in profile. The inner gorge is carved into the Elephant Canyon Formation and lower Cutler Formation, with a bench above formed in the Organ Rock Shale, a finer-grained version of the upper Cutler Formation. The white rock making a prominent cliff under a broad bench is the White Rim Sandstone of Permian age. Rising in a brown slope above the White Rim is the Moenkopi Formation, capped by another bench-forming sandstone, the Moss Back Member of the Chinle Formation. The long slope beneath the massive Wingate cliffs is on the upper Chinle Formation. And finally at the top, the guardian angels of Canyonlands, the Wingate and Kayenta formations, comprise the brown, vertical cliffs. If you look closely in the sharp tributary canyon ahead in the middle distance, the lower reaches of the Shafer Trail dirt road may be seen.

Now wander back through the shelter to the far side of the point for a view to the east. The mountains to the southeast are the La Sals, a laccolithic intrusive igneous range like the Abajos to the south. Highly fractured white rocks of the Navajo Sandstone are visible just below the La Sals along the southwest flank of the Moab salt structure. Lower and closer is a fine view of the Cane Creek anticline, a sharp northwest-trending fold.

Beautifully blue, square evaporation ponds mar the landscape to the west of the anticline. Both halite and potash salts are mined by pumping water down 2,500 feet into the Cane Creek anticline to dissolve rock salt, and the resulting brine is pumped into pits for reprecipitation of the salts. The brilliant blue color of the pits is due to chemicals placed in the brine to hasten

View southeastward from Dead Horse Point of Cane Creek anticline and evaporation ponds. Cane Creek anticline formed by upward arching of the strata by salt flowage at depth. The salt contains potash that is being solution-mined and reprecipitated in the evaporation ponds for use mainly as fertilizer. La Sal Mountains in the distance.

evaporation. The salts are then separated; potash salt is sold for fertilizer and most of the halite is discarded.

Return eight miles to the first intersection. Turn left to Canyonlands National Park and the Visitor Center, a distance of about 12 miles. Access to all points of interest in the Island in the Sky District of Canyonlands National Park is by paved roads, a far cry from the washboards and sand pits of the old road system. So smile when you pay the fee at the park entrance.

GRAND VIEW POINT

The road beyond the park entrance winds around and through rocky knolls of beautifully cross-bedded ancient dunes of the Navajo Sandstone near its contact with the Kayenta Formation. At the first paved intersection, turn left to Grand View Point. And a grand view it is! But watch that first step! You are again standing atop a 600-foot-high vertical cliff held up by the Kayenta Formation and Wingate Sandstone near the southernmost tip of the Island in the Sky.

The Abajo Mountains to the south and the La Sal Mountains to the east serve as landmarks for orientation. The Needles District, with its ragged topography of pinnacles, knobs, and buttes, is more clearly seen here beneath the skyline of the Monument Upwarp. The inner gorge is Meander Canyon of the Colorado River, with its steep walls of the Elephant Canyon Formation. Above the inner gorge are the reddish brown rocks of the Cutler Group, but here and there white beds, tongues of the Cedar Mesa Sandstone, are interspersed. The broad white bench directly below Grand View Point is held up by the White Rim Sandstone at the top of the Permian section. A magnificent example of fracture-controlled pillar and pinnacle topography in the Cutler and White Rim strata lie far beneath your feet in Monument Basin. Then the stairstep slopes of the Moenkopi and Chinle formations rise to the Wingate-Kayenta cliffs. The ruler-straight roads on the bench below that seem to go nowhere were made for seismic surveys in the 1950s in the search for oil, before Canyonlands National Park was even a gleam in anyone's eye. Although the roads were reseeded, they remain obvious scars today.

After a peaceful lunch at the picnic grounds, return to the first intersection. Take a left and then immediately another left turn for a brief trip to Green River Overlook and a pleasant campsite, if you wish.

GREEN RIVER OVERLOOK

Just another lousy overlook in Canyonlands! But here you are looking westward to the Green River far below. The layered rocks look about the same, but the White Rim Sandstone is here obviously thicker than where you've seen it before. Across the river to the southwest is The Maze, a white expanse of highly eroded Cedar Mesa Sandstone, guarded by Elaterite Butte on the right and topped by The Land of Standing Rock. Elaterite Basin, with its tar seeps in the White Rim Sandstone (see Chapter Seven), lies just beyond Elaterite Butte in the sharp tributary canyon. The Orange Cliffs

*Aerial view of White Rim and Monument Basin from above the Colorado River.
Reddish-brown rocks below the White Rim, midway up the cliffs, are in the Cutler
Formation. The distant cliffs consist of the Moenkopi and Chinle Formations in the
slopes, capped by cliffs of the Wingate, Kayenta, and white Navajo sandstones above.*

beyond are the western Wingate-Kayenta bastions of Canyonlands, and the Henry Mountains are the laccolithic peaks in the distant west.

Return through the campground to the first intersection, and turn left to Upheaval Dome.

UPHEAVAL DOME

Park your car at the end of the road and walk about a quarter mile up the steep trail to the overlook into Upheaval Dome. The climb is well worth the effort, but bring along some water if it is hot.

You are looking down into a huge circular depression, almost like a crater, from your vantage point on the Wingate Sandstone. The rocks are dipping strongly away in all directions from a central, very sharp dome. The jagged white pinnacles in the center of the dome are sandstone dikes that squeezed up through fractures from the White Rim Sandstone. Tan-colored and brown shales circling the inner spires are the Moenkopi Formation, overlain by the multicolored Chinle Formation. The massive vertical cliffs are again the Wingate Sandstone that provide your viewpoint. Ledgy beds of the Kayenta Formation form the race-track valley behind you, where your car is parked, and surround the dome. The light colored rocks still beyond, and the farthest from the apex of the structure, are in the old familiar Navajo Sandstone. Like the fins in Arches, an aerial view provides full appreciation.

But what in the world would form such a striking dome? Some geologists believe it is a "cryptovolcano," a volcano that tried to erupt but could not make it to the surface. Others think it is a meteorite impact crater, but there are no meteorite fragments to be found. And those who know salt domes best say it is a classic salt dome, like those in the Gulf Coast region and in Germany. Now it is again thought to be a meteorite impact feature, but of a special kind.

Gene Shoemaker, the brilliant geologist who founded the U.S. Geological Survey lunar geology section for the space program, concluded that a large meteorite impact feature, one considerably larger than Meteor Crater in northern Arizona, should be present on the Colorado Plateau. He decided that Upheaval Dome might be such a feature. He restudied the structure and now believes that it was formed when a meteor hit the Earth back some 65 million years ago when a mile of sedimentary rocks covered the Navajo Sandstone, and before erosion stripped the Colorado Plateau. Because Upheaval Dome has all the earmarks of a salt dome, Shoemaker concluded that some salt flowage was triggered by the meteorite. So where are the meteorite fragments or chunks? The remnants of the meteorite were eroded away along with the mile of sedimentary rocks. He believes that we are looking at the deep-seated core of an impact structure that was a mile below the surface when the thing hit. It would seem that one invariably finds what one is looking for!

Salt will flow upward when anything happens to release the pressure of overlying rocks. There are thousands of salt domes of this type in other parts of the world. And salt flows in circular cells, or mushroom-shaped cells if it can. In the case of the elongate salt structures, such as at Moab and Salt valleys, the salt flowed up along very large and long faults. Yet when the

Aerial view of Upheaval Dome in northern Canyonlands National Park. It is a circular salt-intruded dome, perhaps triggered by the impact of a meteorite. The jagged rocks at the center of the dome are White Rim Sandstone, surrounded by slopes of Moenkopi and Chinle shales. Inner cliffs are the Wingate Sandstone, surrounded by a "race-track valley" of upturned Kayenta sandstones and the outer light-colored, more rounded cliffs of Navajo Sandstone.

fractures around those structures are studied, it turns out they are elongate series of circular salt cells lined up in a row. When salt flows into a vertical subsurface pillar, such as at Upheaval Dome, salt migrates from surrounding areas, leaving a circular depression adjacent to the dome. There is such a "rim syncline" around Upheaval Dome. It is noticeable on the drive from the parking lot near Whale Rock, where limy beds in the Navajo Sandstone dip inward toward the dome. A geophysical study conducted by the U.S.G.S. in the 1960s indicated that there is thickened salt beneath Upheaval Dome. To make matters worse, the same survey showed the presence of a high knob, or something of the sort, in the basement rock directly beneath the structure. It turns out now to be an intersection of two basement faults; such things also trigger salt flowage.

So Upheaval Dome is a true salt dome - a classic beauty of a salt dome. But how was salt flowage triggered? Perhaps it was a meteor that just happened to hit a spot where there was plenty of salt in place and a nice pair of cooperative basement faults beneath. If that is the case, should we rename the thing "Downheaval Dome?"

With this enigmatic structure gnawing at your mind, go back to the car and return to the park entrance.

SHAFER TRAIL

There are two ways to return to Moab. One is to retrace your route by pleasant paved highways. The other is WAHOO! Otherwise known as Shafer Trail. Regardless of your return plans, at least stop at the Shafer Trail Overlook just before reaching the Visitor Center. This is a spectacular view straight down on the myriad switchbacks that lead through the cliffs of Wingate-Kayenta sandstone and beyond. If you are travelling by family auto, you should, by all means, return by paved road. If you are driving a high-clearance vehicle such as a pickup, van, or four-wheel-drive field car, and the roads are not muddy, or you just don't care about tearing up your car and walking back to town, then you owe it to yourself to return via Shafer Trail.

The first turnoff to the right after passing the Visitor Center is Shafer Trail. It is a pretty good dirt road at first, obviously designed to fake out the unwary, but then it goes onto a very skinny ledge of Kayenta sandstone, and then **down** the cliff. The switchbacks are tight, but possible; the road is narrow, but wide enough; it is steep and rocky, and guaranteed to burn out your brakes if you don't use low gear and go very slowly. And one should keep a sharp lookout for uphill traffic, as it is difficult or impossible to pass just anywhere along the way!

For the first couple of switchbacks, the road descends through the ledgy cliffs of the Kayenta Formation. Then the cliff becomes vertical and smooth as you drive downward across the Wingate Sandstone. The cliff finally changes into the steep, ledgy, varicolored slope of the Chinle Formation, and the gradient of the road eases up a little, but very little. A roadcut soon exposes ancient stream deposits of the light-colored Moss Back Member of the Chinle, and descends on through dark brown shales of the Moenkopi Formation. The former Atomic Energy Commission built Shafer Trail from an old stock trail to encourage uranium exploration in the Moss

Back Member in the 1950s boom time. This also explains the many bulldozer scars in the cliffs along exposures of the Moss Back. The trail was no doubt used originally by Indians, then cattlemen, and finally uranium prospectors.

About where the road finally flattens, there is a large turnout on the right. You should stop here to let the smoke out of your brakes, and get out to breath a sigh of relief and stretch your shaky legs. Take a look over the edge of the turnout, a good excuse for stopping, where 'dozers have pushed big slabs of rock out of the way. Bedding planes on these rocks reveal excellent examples of several kinds of ripple marks, mud cracks, worm burrows, and load features in the Moenkopi Formation. These "sedimentary structures" tell us that the formation was deposited as a broad tidal flat in Lower Triassic time, some 245 million years ago last Tuesday.

After the brakes have cooled, go on down the hill a short distance to a fork in the road. Straight ahead is the White Rim Trail that snakes around innumerable tributary canyons for about two days at the top of the White Rim Sandstone and the base of the Moenkopi Formation. It eventually comes out along the Green River to Mineral Bottom, leaving the canyon for Moab via the Horsethief Trail, another cute little road up the cliff. This is really the hard way to get to town! One truly needs four-wheel-drive to go very far on this road.

Fortunately, there is another way, and that is to turn left. The rough road immediately descends through the White Rim Sandstone. The upper half of the formation is highly cross bedded, indicating that it was deposited as windblown dune sand. The lower half is flat, crinkly bedded, suggesting deposition on a sandy tidal flat. Reddish brown rocks below the White Rim are mostly stream-deposited feldspar-rich sandstones (arkose) of the Cutler Formation. You might see some bighorn sheep in this area.

The trail bottoms-out where Shafer Canyon challenges the uppermost limestone bed of the Elephant Canyon Formation. The gray, hard limestone, locally called the "Shafer lime," is dipping steeply toward the north on the flank of Shafer Dome, an east-trending anticline that is pillowed, but not pierced, by salt flowage. About 5,000 feet of Paradox salt underlies the sharp fold (see Chapter Six).

The road follows the top of the Elephant Canyon limestone for several miles along the flank of Shafer Dome. Red sandstones of the Cutler Formation form the cliff on the left. The limestone contains lots of fossil clams, snails, brachiopods, a rare trilobite or two, and other miscellaneous goodies. Large fossil clams make up the road gravel in places. Views into the beautiful inner gorge of the Colorado River are common. Dead Horse Point, with the large arch-shaped cove in the cliff and the shelter that looks like a giant hang glider, is occasionally visible high on the left.

After the road descends the steep flank of Shafer Dome for the last time, evaporation ponds as seen from Dead Horse Point finally appear and disrupt the route of travel. They are lined with heavy black plastic to control seepage, although this is obviously not one hundred percent effective. Brine is pumped through pipelines from the solution mine, some 2,500 feet beneath the Cane Creek anticline, into the ponds, where evaporation reprecipitates the salts. Both potash and halite salts are harvested with large earth-moving vehicles, and hauled to the large green structure at Potash for separation.

White Rim, near Monument Basin. The ever-pillared margin of the White Rim bench is here forming before your eyes. Erosion widens fractures faster in the softer reddish-brown Organ Rock Shale than in the more resistant White Rim Sandstone above, forming the pedestals.

POTASH

The large, green, domed buildings at Potash are product warehouses of the Moab Salt (owned jointly by Texasgulf International and Moab Salt companies) potash mine. A shaft goes down 2,200 to 3,200 feet, and then a 2,500 foot tunnel enters the potash-rich salt body beneath the crest of the Cane Creek anticline. Salt was originally mined with heavy equipment, starting in late 1964, but conventional mining was soon found to be impractical. It was hot in the mine, about 100 degrees Fahrenheit, and gassy. There was an explosion in the mine that killed 18 workers. The salt beds were highly contorted due to salt flowage, making the use of earth movers difficult.

The company converted to solution mining in 1970. The first harvest of salt under the evaporation system occurred in March, 1972. Five wells were drilled into the structure for water injection, and flooding with river water began on Christmas Day, 1970. It takes about 300 days for the injected fresh water to travel through the mine, become saturated with dissolved salts, and be pumped out to the 23 evaporation ponds.

Potash salt is separated from the halite (table salt) by a flotation process, and shipped from the mine by rail. The mine is expected to produce about 300,000 tons of potash, mainly used for fertilizer, during its 20-year projected life span. Some of the halite is sold for use on Utah highways, but otherwise there is only a very limited market for table salt in Moab. Some has been sold for oil well drilling operations, for salt blocks for cattle, and other uses. Excess halite is stored behind an earth-fill dam in the tributary canyon behind Potash; company officials say it is safe from flash-flooding.

Leaving the evaporation-pond farm, the road descends sharply through a thick purplish-colored bed of windblown Cutler sandstone, and then crosses the Cane Creek anticline. This is another salt-pillowed fold, but unlike Shafer Dome, trends northwesterly paralleling the Moab salt structure. Thin limestone beds of the Elephant Canyon Formation (Permian Period) cap the anticline.

Several six-foot-tall pipes may be seen near the crest of the anticline. These are markers for old abandoned oil wells drilled back in the good old days by Monty G. Mason (M.G.M. Oil Company). Rumors linger that Mason made shady financial deals regarding his "oil play," but Bob Norman, a longtime consulting geologist in Moab tells this story.

"After putting together a block of oil properties at the Cane Creek anticline in 1948, Monty G. Mason raised sufficient capital to drill his first test for oil and gas along the Colorado River bank at the crest of the anticline. This location, the Americol Petroleum Mason No. 1, was within 50 feet of two holes drilled in 1925 and 1940; Midwest Exploration Cane Creek Gusher, and Cane Creek Petroleum No. 1 Shafer respectively. The 1925 test hole encountered oil, caught fire and burned completely. Although some oil shows

Aerial view of the Cane Creek anticline north of the Colorado River. Gray rocks that form the core of the up-folded structure are limestone beds of the Elephant Canyon Formation. Red cliffs that ring the fold are in the Cutler Formation. The far slopes represent the Moenkopi and Chinle Formations, and the far cliffs are the Wingate and Kayenta formations. Buildings in the upper far right of the photograph are at Potash.

were encountered, the Mason No. 1 test was eventually plugged and abandoned.

"On October 30, 1953, Monty Mason moved to a new location farther up the hill along the anticlinal axis and away from the river. His second test was called MGM Petroleum MGM No. 1. Rotary drilling was carried to a depth of 7835 feet within the base of the Paradox salt. No commercial quantities of oil or gas were obtained at this site.

"On December 1, 1954, Mason moved to a third location between Mason No. 1 and MGM No. 1. This test was drilled to a depth of 7495 feet and tested an excellent oil show at 6738 feet within the "non-salt" zone known as the 'Cane Creek marker.' High gravity oil flows were encountered on a series of tests and, with a 16 mm camera, many of the flows of oil and some gas, which was flared, were filmed. Monty Mason stayed with the operation until 1956. Even though he raised considerable money for continued testing, he was never successful in maintaining a sustained flow of oil.

"The State of California prosecuted Monty Mason for overselling stock in several corporations which controlled overlapping mineral rights. He spent several years in a California prison, and being a diabetic, passed away in 1965 prior to his eligibility for parole.

"Most of Monty Mason's business associates and the citizens of Moab thought of him as a friendly, good-hearted individual. He maintained an office downtown with a fixed, flashing, red neon sign reading 'Oil Operator.' He successfully raised money elsewhere and freely spent the cash in Moab helping to keep Moab's economy at a high level until the beginning of the uranium boom in 1952."

Thus, a highly exaggerated rumor is clarified.

Now the road descends the eastern flank of the Cane Creek anticline on bumpy bedding planes of fossiliferous limestones. As it reaches river level, a launching site for Cataract Canyon river trips is visible on the right. Roads and pipelines go everywhere, but the "correct" road is usually recognizable. Finally, the rough road changes to a paved highway leading from the Potash processing plant (large, green structure to the left) to Moab. It feels like you have just bought a new car, as you head for town on, at last, pavement. From Potash to Moab, the highway crosses the King's Bottom syncline that separates the Cane Creek and Moab anticlines. The magnificent cliffs along the river are mostly in the Glen Canyon Group of Jurassic age.

CHAPTER 13
THE NEEDLES DISTRICT

EN ROUTE

To visit the Needles District of Canyonlands National Park, head southeast from Moab on U.S. Highway 191 toward Monticello, along the length of the Moab Valley salt anticline. The highly fractured and slumped cliffs on either side of the broad-floored valley consist of the white Navajo Sandstone, overlying the brown, ledgy cliffs of the Kayenta Formation. Dark brown, massive cliffs below are in the Wingate Sandstone. Slopes of Chinle shale are occasionally visible just above the valley floor, beneath the shattered and collapsed Wingate cliffs. The La Sal Mountains dominate the skyline, ahead and to the left. They are an intrusive igneous "laccolithic" range that rises to a height of 12,721 feet at the summit of Mount Peale, the fifth highest peak in Utah.

The hummocky hills ahead and to the left in the valley floor, 10 or 12 miles south of Moab, are jumbled layers of the light-colored Morrison Formation and the dark gray, weathered to sickly yellow, Mancos Shale, that have collapsed into the end of the salt structure. As the highway takes a right turn and starts up a hill, complete with passing lane, the strata ahead first dip noticeably northward toward the valley at the hinge of the fold, where younger strata are drooping into the collapsed top of the salt anticline. Then the beds flatten and begin to dip away from the valley, reflecting the original upward arching of the saltpierced structure.

The highway is then on alternating sandstones and shales of the lower Morrison Formation (Salt Wash Member) as it crosses the high flats and begins its descent into a sharp canyon. On the left, a well-exposed small fault displaces the beds of Morrison Formation against the sheer, red cliffs of the Slickrock Member of the Entrada Sandstone. The Slickrock Member then forms the walls of the beautiful oasis of Kane Springs, and hosts the outrageous announcement regarding "Hole-in-the-Rock." Do not confuse this establishment with the historical crossing of Glen Canyon by the early Mormon pioneers!

From Kane Springs to the Canyonlands Junction, about 33 miles south of Moab and 20 miles north of Monticello, the highway wanders to and fro among exposures of the Navajo Sandstone below, crinkly brown siltstones of the Carmel Formation (Dewey Bridge Member) in roadcuts, and the spectacular red cliffs of the Entrada Sandstone above the highway. And watch for the strong and youthful Wilson Arch on the left in the Slickrock Member of the Entrada Sandstone.

An optional interlude is to turn right (west) at the Needles Overlook Junction for spectacular views into the southern parts of Canyonlands. For several miles the paved road generally traverses flatlands in and near the base of the Navajo Sandstone. At Needles Overlook one is treated to a magnificent overview of the Needles District, with all of its "ten thousand strangely carved forms." Besides an aerial view of the shattered landscape below, you can see the interesting pattern of the red Cutler sandstones and the white Cedar Mesa Sandstone from the west. Note that the low country west of Cataract Canyon is of white rock, but the land below is red. The two rock types interfinger, forming the banded appearance in the Needles (see Chapter Seven). If time permits, the drive to Anticline Overlook is worth the effort. The dominant view is of the Cane Creek anticline, a grand northwest-trending upfold, and a look back across the Colorado River at Dead Horse Point.

Meanwhile, back on U.S. Highway 191 heading south, take a right turn to the west at Church Rock (Entrada Sandstone) and head for Newspaper Rock State Park and the Needles District of Canyonlands National Park.

Ute Indians ambushed a field party of the Hayden Survey at the base of the hills just southeast of Church Rock on August 17, 1875, but the surveyors managed to escape. Their abandoned equipment and spent cartridge cases were found by P.K. Hurlbut in 1964.

The prominent range to the south is the Abajo Mountains, or "Blue Mountains" as they are known locally. In the 1700s, the Spanish padres named them Abajo ("lower") Mountains as compared to the La Sals farther north. The summit is Abajo Peak at an elevation of 11,360 feet. Like the La Sals, the Abajos are eroded remnants of intrusive igneous rock bodies, or "laccolithic" mountains. Farm buildings along the road are abandoned remnants of an old religious commune, known as the "Home of Truth," founded by Marie Ogden.

After passing through a notch, the highway winds down into a sharp canyon, and the rocks start looking peculiar. This is the Shay Graben, a down-faulted block. The cliffs to the left are the Navajo-Kayenta-Wingate sandstones in descending order. Only the Navajo is seen right of the fault, however, as it is faulted down adjacent to the Wingate Sandstone. Then just before reaching Newspaper Rock, a second fault is crossed and the Wingate comes back up to the surface. Newspaper Rock displays graffiti of the past thousand years or so, carved into desert varnish in an alcove of Wingate Sandstone. Desert varnish is a surface coating of iron and manganese oxides; origin of the phenomenon is debatable and basically unknown.

Heading west down Indian Creek, the Wingate cliffs become progressively higher and more spectacular. But why is the Wingate a white color here, instead of the usual brown? Red and brown coloration is due to the presence of iron minerals in the oxidized, rather than reduced, state. Either the iron minerals are lacking in this area, or they have undergone chemical reduction, in other words, have been bleached. Your guess is as good as any as to why or how this happened. About five miles from Newspaper Rock, the varicolored slopes of the Chinle Formation rise to the surface and the typical Canyonlands facade is again recognizable.

At Dugout Ranch the canyon broadens, as the valley floor is now on the more resistant Moss Back Member of the Chinle Formation, a stream deposited pebbly sandstone. This was the center of a giant cattle empire in the early 1900s. Dugout Ranch was owned by the Scorup and Somerville brothers after 1918, and the Scorup cattle company became the largest outfit

Aerial view looking south into the Needles District of Canyonlands National Park. Red and white banded cliffs in the foreground are an especially well exposed "facies change" formed by the mixing of two sediment types. Weathering along the intricate fractures carved the pinnacles and toadstool rocks. The gently arched skyline is the crest of the Monument Upwarp in profile. Navajo Mountain may be seen in the far distance.

in Utah and the second biggest in the United States by 1940. As many as 15,000 head of cattle were driven annually through Moab to the railroad at Thompson. Thousands of square miles of summer grazing lands lay a few miles to the south on Elk Ridge, and Indian Creek made fine winter pasture land. What more could a cattle baron need? It is still a working ranch.

About five miles past the ranch turnoff, the highway breaks over the erosional edge of the Moss Back bench, and drops down through the dark brown mudstones of the Moenkopi Formation. North and South Six-Shooter Peaks, capped by the Wingate Sandstone, are to the left. Red slopes are in the Organ Rock Shale of Permian age, and the highway crosses onto the arkosic sandstones of the lower Cutler Group (Lower Permian) after crossing Indian Creek. Note that the White Rim Sandstone, so prominent below the Island in the Sky, is missing here. It pinches out toward the east along the course of the Colorado River. When The Needles come into sight, the banded red and white coloration is due to interfingering of the white Cedar Mesa Sandstone with tongues of red Cutler arkose (see Chapter Seven for further explanations).

THE NEEDLES

If the layered rocks of the Needles District appear to be highly fractured like a sheet of broken glass, it's because they are. Vertical fractures (joints) are closely spaced in a northeast and northwest criss-crossing pattern. Erosion has widened the fractures, leaving rock remnants standing as giant toadstools and pinnacles: The "ten thousand strangely carved forms" of John Wesley Powell. The toadstool shape results from the white tongues of Cedar Mesa Sandstone being more resistant to erosion than the red tongues of Cutler. Where the white rock caps the red, the softer red sandstone weathers back to form a base of the pedestal.

Why are the rocks so highly shattered here? For one thing, the Needles District lies astride the crest of the very broad, plunging fold of the Monument Upwarp. Although such an upfold in the rocks was caused by compression in the crust, the crest of the fold is under tension and the layered rocks want to crack apart and collapse. Another factor is that the area is underlain at rather shallow depths by salt of the Paradox Formation. Near-surface groundwater trickling down through overlying rocks dissolves some of the salt, allowing the overlying strata to settle or collapse, causing further fracturing or opening of existing fractures. Because the fractures are closely spaced and are formed in opposing directions, spires and "toadstools" result, instead of "fins" as in Arches National Park.

The scale of fracturing is magnified greatly as the rock layers cross the crest of the huge fold and dip westward toward the Colorado River. The river has carved a huge 2,000-foot-deep canyon into the rocks, nearly as deep as the salt itself, in effect undercutting the highly fractured rocks. The broken blocks are actively slumping and sliding down into Cataract Canyon on the slippery salt layer, opening up the fractures, with some blocks collapsing along faults in the process. The result is the area west of The Needles known as "The Grabens." A graben is a down-faulted block of rock; a horst is an up-faulted block. And horsts and grabens are a dime-a-dozen

A whole field of "toadstool rocks" in the Needles District of Canyonlands National Park. Weathering of the rock layers of different hardness along numerous fractures form the novel topography.

Geological illustration showing the formation of the Grabens in Canyonlands National Park. The Colorado River has carved Cataract Canyon to within a few feet of the top of the Paradox salt beds (IPps), literally undercutting the inclined layers of the overlying Honaker Trail (IPht), Elephant Canyon (Pec), and Cedar Mesa (Pcm) formations. The rock layers above the salt have broken up and are sliding toward the canyon on the top of the relatively slippery salt. The course of the river follows the crest of Meander anticline, a salt-intruded fold located above a deep-seated fault zone.

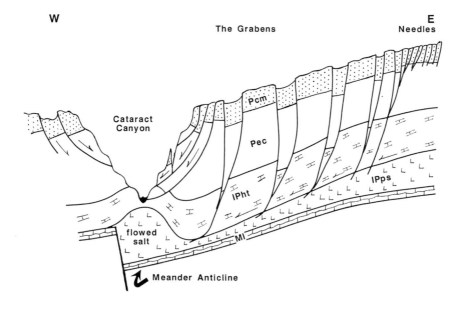

Aerial view of Cataract Canyon and the Grabens. The confluence of the Green and Colorado rivers is barely visible in the lower right corner of the photograph. The Colorado River disappears into Cataract Canyon at Spanish Bottom (Cataract Bottom) at right-center. Obviously shattered rocks in the left half of the view constitute the Grabens of the Needles District of Canyonlands National Park. The cliff-forming caprock is the Cedar Mesa Sandstone. Navajo Mountain is on the left horizon and the Henry Mountains are visible on the right.

Block diagram showing the most prominent topographic features in and near Canyonlands National Park. Notice that the entire area is underlain by salt in the Paradox Formation. The location of the Needles is near the crest of the Monument Upwarp, and the Grabens have formed where rock layers above the salt have broken apart and are sliding down the west flank of the fold to the canyon. (Drawing by William L. Chesser, courtesy of Canyonlands National Park)

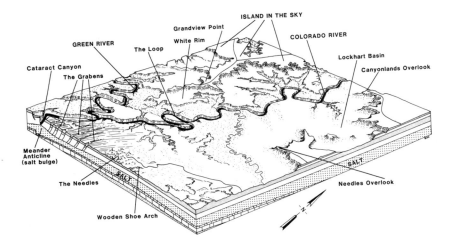

between the Squaw Flats Campground and Cataract Canyon. The giant slumping grabens are actively widening today, causing measurable opening between fault blocks, even as we speak. Routes of the four-wheel-drive trails west from the Squaw Flats Campground over Elephant Hill into Devil's Kitchen and Chesler Park country are restricted by this web of complex horst and graben topography. Because the grabens are still actively collapsing, care must be taken not to walk or drive near the faults, as the ground often caves into the open voids. These faults occur immediately adjacent to the many long, nearly vertical cliff faces.

The several arches in southern Canyonlands National Park are formed in the Cedar Mesa Sandstone. Fracturing is not nearly so well developed in that area, but erosion has carved an intricate pattern of small canyons into the Cedar Mesa surface. Where canyons are closely spaced, or where one doubles back on itself to form a "gooseneck," it is easy for the massive rock to spall and break through a ridge, forming a natural arch. There are more than a dozen arches in the intricate drainage system of upper Salt Creek. Four-wheel-drive trails south from the Park Entrance wander about through this complexly eroded maze.

Do not attempt to drive a street vehicle much beyond the Squaw Flats Campground in any direction. The roads, either into The Grabens or south to the various arches, are strictly for high-clearance, four-wheel-drive vehicles only. Several guide services are available in Moab, Monticello, and near the park entrance to make such trips safe and much more enjoyable.

The rock column at and near the confluence of the Green and Colorado rivers in Canyonlands National Park. Rocks of the Paradox, through Cedar Mesa formations form the cliffs of Cataract Canyon; overlying formations have been stripped back by erosion to form the outer canyon walls.

GREEN & COLORADO RIVERS
STRATIGRAPHIC COLUMN

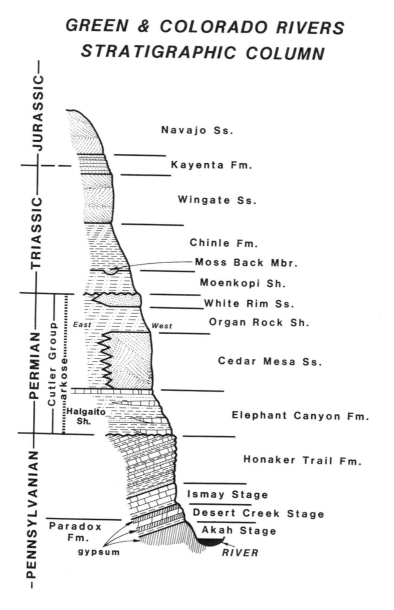

Navajo Ss.

Kayenta Fm.

Wingate Ss.

Chinle Fm.

Moss Back Mbr.

Moenkopi Sh.

White Rim Ss.

Organ Rock Sh.

Cedar Mesa Ss.

Elephant Canyon Fm.

Honaker Trail Fm.

Ismay Stage

Desert Creek Stage

Akah Stage

RIVER

Paradox Fm.

gypsum

Halgaito Sh.

Cutler Group
arkose

East   West

JURASSIC

TRIASSIC

PERMIAN

PENNSYLVANIAN

CHAPTER 14
THE MAZE/ELATERITE BASIN

The west side of Canyonlands National Park and northern Glen Canyon National Recreation Area are fascinating areas, both scenically and geologically. But, as they say, you can't get there from here. The best route is to go west from Blanding to Hite Marina on Lake Powell via Utah Highway 95, or south and east from Hanksville to Hite on the west end of the same highway. There are two bridges on Highway 95 near Hite, one over the Colorado River canyon and the other across the Dirty Devil River. An obscure dirt road turns north from the highway between the two bridges, opposite Hite Airport. It is about 40 or 50 miles to any destination this way, so settle down to a long, rough drive. Only four-wheel-drive vehicles should be used! Take plenty of water, a sack of sandwiches, and gasoline, as there are no McDonalds on this route. A tool kit and a survival kit are always in order, and a CB (citizen's band radio) is desirable if travelling alone in the back country.

There is another way to approach The Maze District. About half way between Green River and Hanksville, a dirt road goes east from Utah Highway 24 towards Canyonlands National Park. There is about 55 miles of rough road leading to the head of Flint Trail. If you liked Shafer Trail, you'll love Flint Trail. It is an old cattle trail that was improved by the Atomic Energy Commission in the early 1950s to encourage uranium prospecting west of the Colorado and Green rivers. Flint Trail is steep, narrow, and rocky, and has tight switchbacks. Four-wheel-drive and a strong heart are required!

EN ROUTE VIA BLANDING/HITE

Four miles south of Blanding, turn right (west) on State Highway 95. The road goes down through the several formations of Jurassic age until it reaches the white, massive, cross bedded Navajo Sandstone. By this time you will notice that the beds are turned up sharply toward the west along the Comb Ridge monocline. Before very far the highway turns westward and crosses the sharp fold in the rocks, first passing through a roadcut in the Navajo-Kayenta formations, and then descending through the Wingate Sandstone (watch for rockfalls here) and the Chinle Formation into Comb Wash. This is the site of the bulldozer massacre by the "Monkey Wrench Gang," heros in the novel by that name written by Edward Abbey.

Then the highway climbs rapidly up the east-dipping limb of the monocline on the Cedar Mesa Sandstone of Permian age. (For an explanation of monoclines, see Chapter Two.) The roads will follow the top of the Cedar Mesa Sandstone, along the base of the Organ Rock Shale, the rest of the trip across the Monument Upwarp to Hite and north to The Maze. This makes a good highway route because the sandstone is very resistant to erosion and forms broad benches and plateaus, while the much softer Organ Rock red beds form smooth road surfaces. The broad plateau to the south of the highway, after reaching the high country, is Cedar Mesa, for which the sandstone is named. The two prominent buttes on the skyline to the north are the Bear's Ears that rise to 9,058 feet; they are capped by the Wingate Sandstone.

Be sure to stop at Natural Bridges National Monument, as it is only a few miles from Highway 95. A contemporary Visitor Center, powered entirely by solar energy, contains exhibits that explain the formation of the natural bridges that are totally unlike arches. Incised stream meanders in White Canyon have been eroded completely through at their goosenecks, allowing runoff waters to cut off the meander loops and flow under remnant rock bridges of Cedar Mesa Sandstone. A short loop road allows easy access to the three natural bridges in the park, and a pleasant campground.

From Natural Bridges, the highway starts down the long, gentle west flank of the Monument Upwarp, along White Canyon to Hite. White Canyon was one of the largest uranium mining districts, and its legacy is evident along the canyon walls. Production was from the Shinarump Member, the lowest bench-forming layer of the Chinle Formation. The next bench-forming sandstone above the Shinarump is the Moss Back Member that is widespread in Canyonlands. Fossil plant material in the ancient stream-deposited sandstone created conditions in the rock that precipitated uranium and other minerals from groundwater.

The white rocks of the valley floor are still Cedar Mesa Sandstone. Reddish brown slopes from road level up to the mining horizon are in the Organ Rock Shale (Permian Period), and the Moenkopi Formation (Triassic Period) in the upper slope. The White Rim Sandstone first appears in the section between the Organ Rock and Moenkopi red beds in the low cliffs and buttes near Jacob's Chair. The flat-bedded sandstone was here deposited on a beach, and pinches out toward the east. The higher cliffs and buttes along the margins of White Canyon are capped by the Kayenta-Wingate Sandstones. The time-rock charts inside the book covers help organize the rock and age sequences.

Although one needs an eagle eye to notice it, the entire route from Comb Ridge to Hite, and beyond up North Wash, is along the Four Corners Lineament. This deep-seated fault zone has only faint surface expression. White sandstone dikes cutting upward through fractures in the Organ Rock Shale just east of Hite, and the eroded fracture zone seen on the skyline northwest of Hite, reflect this ancient fault zone.

Turn left to Hite Marina. Fill gas tank and spare cans, buy ice cream, candy bars, soda pop, and other goodies, and pick up a shaker of salt for the one you forgot to put in the food box. This is your last chance for such conveniences and necessities. Then head back west across the Colorado River (now Lake Powell) bridge, and turn right (north) on the unmarked dirt road across from the airport. If you get as far as the bridge over the Dirty Devil River (now Lake Powell) you have gone too far; turn back for the dirt road.

It is 46 miles to Canyonlands National Park, 37 miles to the foot of Flint Trail, 19 miles to Cove Springs, and 45 miles to Elaterite Basin over

Bridge over the Colorado River at Narrow Canyon near Hite Marina. Narrow Canyon is now inundated by Lake Powell. The lower, light-colored cliffs are eroded from the Cedar Mesa Sandstone of Permian age (labelled Pcm), overlain by the darker Organ Rock Shale (Por). The White Rim Sandstone (Pwr) and Triassic Moenkopi Formation (Trm) cap the butte.

bone-jarring washboards and slickrock roadways. The road is along the Cedar Mesa Sandstone-Organ Rock Shale contact most of the way, with prominent white cliffs of the White Rim Sandstone above the lower red beds. The high cliffs, called the Orange Cliffs, and the higher buttes, such as the Sewing Machine, are capped by the Wingate-Kayenta Formations.

It is wise to stay on the main road, which is obvious, and not stray onto unmarked side roads, or try "bushwhacking." Deep sand and rough slickrock can be dangerous to your vehicle's health. About the time you think that the road is endless, Teapot Rock, capped by the Moss Back Member, looms ahead, and a marked road goes right toward The Maze and the Land of Standing Rocks. The road straight ahead goes to Elaterite Basin and the Maze Overlook.

THE MAZE

It is only about 10 miles as the crow flies from Teapot Rock junction to the heart of The Maze, but it seems more like 10 days by jeep trail. The road follows the topographic bench at or near the top of the Cedar Mesa Sandstone all the way. The main road destination is the Land of Standing Rocks, where buttes and spires of Organ Rock Shale rise above the top of the Cedar Mesa Sandstone. Spectacular erosional remnants, The Wall, Lizard Rock, The Plug, Standing Rock, and Chimney Rock, guard the divide between intricately carved drainage systems that form The Maze to the north and Ernie's Country to the south. There is nothing geologically significant about the elaborate system of canyons surrounding the Standing Rocks, except that the massive sandstone surface has been ravaged by eons of rainwater runoff and its relentless erosional powers.

There are several primitive campsites in the Standing Rocks area, and miles of hiking trails into the tortuous canyon system. Be sure to take along a trail map and plenty of water for venturing beyond the road. A jeep road does lead to the Doll House overlooking Cataract Canyon, but a road indicated on older topographic maps leading to Pete's Mesa to the north is nonexistent. Similarly, don't expect to take a shortcut into Elaterite Basin by car over the "Golden Stairs." That is strictly a foot trail, in spite of map indications. The newer waterproof maps by Trails Illustrated, available at Visitor Centers, are recommended for navigation in and near Canyonlands National Park. Return the way you came to Teapot Rock.

ELATERITE BASIN

Four-wheel-drive vehicles are necessary for travelling the trail from Teapot Rock into Elaterite Basin. You can get in with high-clearance two-wheel-drive vehicles, but you may not be able to get back out.

The road is on the White Rim Sandstone, here saturated with tar, as it rounds the base of Teapot Rock. It then ascends steep slopes of Moenkopi shale and abruptly climbs a rocky route to the top of the Moss Back Member of the Chinle Formation, where large, black sections of petrified trees are weathering from the shale. The road then stays on or near the top of the

Moss Back to Flint Flat. Bentonitic shales of the lower Chinle Formation make this stretch of road a treacherous cliff-hanger when wet. The White Rim Sandstone may be seen to thin from about 250 feet to a pinchout along the north wall of Range Canyon below Flint Flat.

Flint Trail is a shortcut leading from Flint Flat to Utah Highway 24 via the Orange Cliffs. It is a tortuous jeep trail up the Chinle-Wingate-Kayenta cliffs into Robber's Roost country. A four-wheel-drive vehicle is an absolute necessity.

A thickened trend in the White Rim Sandstone extends from Teapot Rock on the south, across Range Canyon, and the length of Elaterite Basin to the north. The barlike sand body is saturated with tar throughout its length. If we could have drilled this feature 50 or 60 million years ago, before erosion exposed the White Rim to the elements, we would have had a giant oil field. The Utah Geological and Mineral Survey estimates that there are 16 billion barrels of oil in place here. As it is, the deposit is a classic example of an exhumed oil trap laid bare for study.

Follow the jeep trail north from Flint Flat, down a steep descent across the Moss Back Member and Moenkopi Formation. It is wise to examine the road ahead before starting down the steep Moss Back cliff, as it is very rocky and may be difficult, or perhaps impossible to get back up on the return, especially when wet. And don't forget, there's no other way out of Elaterite Basin!

After the descent the road follows a wash that parallels a mound of White Rim Sandstone on the right. Although the road is at or near the base of the Moenkopi Formation, the older White Rim Sandstone rises 50 feet or more above the wash to the right. There are small washes that cut natural cross sections through the mound of sandstone, showing that it is a "fossil" sand bar in the White Rim. A short walk into the washes reveals that the lower cliffs are highly cross bedded, while the upper veneer of sandstone is horizontal to gently inclined beds with huge, symmetrical ripple marks on the bedding surfaces. This rippled veneer crosses the entire width of the mound, clearly indicating the original, preserved shape of the sand pile at the time it was deposited.

The washes also reveal large, drippy tar seeps along the cliffs, and the "white" sandstone is found to be black and saturated with tar when broken open with a hammer. In 1964, after a cloudburst that nearly wiped out Flint Trail, these washes were bared of sand by the runoff and were found to be literally "paved" with natural tar seeps (there were no yellow stripes).

The road follows the margin of the "sand bar" for a mile or so, and then climbs to the top and crosses the bar. On the east side, the road descends the preserved slope of the rippled surface. Erosion has exposed the external shape of the bar just as it was in Permian time, some 265 million years ago, when the sea retreated from Canyonlands Country. Perhaps a hundred yards past the flank of the bar, stop in a turnout and walk a hundred feet or so into a small gully to the right. The rocks in the head of the gully are covered with thick, flowed tar from a seep that is still active during the hot months.

Oil is buoyant on water, and all buried rocks below the water table contain water. Any oil present in the rocks will rise to the top of the bed and then travel up-dip within the bed as far as it can go. In this case, oil migrated up through the White Rim Sandstone toward the Monument Upwarp from the Henry Basin to the west. When it reached the up-dip pinchout of the porous sandstone, it had no place to go and was trapped. Oil piled up in the thickened bar, forming a "stratigraphic trap." The oil is now weathering to tar, as it dries up in the desert air.

The presence of petroleum in the White Rim Sandstone led to the bleaching of overlying and underlying red beds. It is especially noticeable above and below the White Rim in Elaterite Basin. Red coloration in rocks is due to the presence of iron in the rocks in the "oxidized" form (Fe_2O_3). In other words, these red minerals contain a high ratio of oxygen to iron due to having been exposed to excess oxygen (they rusted) at some time early in their history. If that ratio is lowered by removing some of the oxygen, a gray or green mineral results (FeO). The presence of petroleum, or any organic matter in the rock, creates a "reducing environment" and red minerals change to gray or green minerals. When the oil-laden White Rim Sandstone was deeply buried before erosion, some of the lighter fractions of the petroleum (probably methane, ethane, etc.) leaked a short distance into adjacent red beds, a reducing environment was created in the rock (in the groundwater), and the affected red rocks were bleached to a tawny color.

This same reducing process forms small, usually silver-dollar-size, green spots in red rocks where a piece of fossil plant material creates a local reducing environment, bleaching the nearby rock. (I'll bet you always wondered why red rocks have green spots!) The same reducing process also localizes uranium minerals around fossil plant debris in the Moss Back and Salt Wash Members. (Look in any uranium mine or prospect pit and you'll find a lot of fossil twigs, leaves, etc.)

What was the origin of the White Rim "sand bar"? We think it is a preserved offshore bar, formed at the margin of an extensive seaway that lay to the west in Permian time. However, folks who think all large-scale cross bedding must have formed in windblown dunes, believe the lower part of the "bar" is eolian. No one disagrees that the upper surface of the bar was rippled in shallow water, as winds do not form large oscillation ripples. Either way, it was a shallow marine bar, at least for a brief interval of Permian history. And it is an excellent trap for oil, even though it is now exposed at the surface and economically worthless.

These exposures in and near Elaterite Basin are basic to the more regional interpretations discussed in Chapter Seven.

The next main road to the right leads to The Maze Overlook, or the road ahead continues for several miles to a dead end at the Green River. The only way out is to retrace the approach route.

CHAPTER 15
LABYRINTH-STILLWATER CANYONS

Cataract Canyon below the Confluence of the Green and Colorado rivers is the heart and soul of Canyonlands Country. Entry to the deepest canyon is via either of the rivers, but it is otherwise almost inaccessible. The expeditions of John Wesley Powell in 1869, and again in 1871, entered Cataract Canyon from the Green River, and named the canyons Labyrinth and Stillwater for their most memorable characteristics. Although there is really only one canyon below Green River, Utah, Powell's propensity for naming canyons compelled him to name more than one in 120 miles of river. Consequently, he changed canyon names where the river crosses from rocks of Mesozoic age to those of Paleozoic age at Mile 36 (measured upstream from the confluence) between Fort Bottom and Anderson Bottom.

Geological characteristics of the canyons will be treated here only in generalities, as the details are spelled out in the waterproof river runners' guide Cataract Canyon and Approaches, published in 1987 by Cañon Publishers Ltd. That field guide is highly recommended for the interested river runner

If one enters Canyonlands Country from Green River, Utah, there is 120 miles of slow river before reaching the confluence with the Colorado River, and Cataract Canyon just downstream. Indeed, the gradient of the Green River is here only about 1.25 feet per mile, and there are no rapids, no riffles along the way. Rocks at the surface around the town of Green River are the drab, dark gray beds of the Mancos Shale of Late Cretaceous age. However, within a mile downriver the Dakota Sandstone, also Late Cretaceous age, comes to the surface. From there the river cuts slowly but surely downsection through rocks of Jurassic age. A small graben, or down-faulted block, crosses the river at Crystal Geyser, which is an unplugged well drilled for oil in 1936. Otherwise, the river cuts its course persistently down through the layers, one formation at a time without structural interruption, until it reaches the confluence. The layered rocks are dipping ever so gently northward, toward the Uinta Basin, while the river is flowing nearly as gently in the opposite direction. This accounts for the appearance of older rocks down-river, and helps slow the gradient of the river at the same time.

LABYRINTH CANYON

At first the Green River flows along an open valley, because the rocks are relatively soft, weathering easily back from the river. When the more resistant rocks of the Glen Canyon Group, the Navajo-Kayenta-Wingate sequence, rise at the river's edge, a canyon develops ever so slowly. The

West wall of Labyrinth Canyon on the Green River near Mineral Bottom. The lower slopes beyond the willow-tamarisk thicket are formed on the upper Moenkopi Formation, the two intermediate cliffs separated by slopes are in the Chinle Formation, and the high cliffs and spires consist of the Wingate Sandstone, capped by thin remnants of the Kayenta Formation.

official head of Labyrinth Canyon is where the Navajo Sandstone first becomes apparent (Mile 94.4). The canyon deepens slowly as older formations rise to the surface, until at Mineral Bottom (Mile 52.2) where there is a boat-launching ramp, the Triassic Moenkopi Formation is at river level in a respectable canyon.

Many Cataract Canyon river trips launch at Mineral Bottom. The access road is the Horsethief Trail, built in the 1950s and maintained for hauling uranium ores from mines in the Moss Back Member of the Chinle Formation. The dark brown shales at river level for the next 15 miles are in the Moenkopi Formation. The low ledges are the Moss Back Member, and the overlying varicolored slopes are upper members of the Chinle Formation. The massive brown cliffs high in the canyon walls are the Wingate Sandstone, capped by the ledgy beds of the lower Kayenta Formation. The Navajo Sandstone is occasionally seen as white domes and rounded cliffs at the canyon rim. The Wingate-Kayenta-Navajo Formations are now considered to be of Jurassic age; the Moenkopi and Chinle are Triassic. Major Powell named this area "Tower Park" during his 1869 expedition, a name that does not seem to have survived.

The dirt road that follows the left bank of the river in this area is the White Rim Trail. It follows the top of the White Rim Sandstone from Mineral Bottom and the Horsethief Trail around the base of the Island in the Sky, eventually connecting with Shafer Trail below Dead Horse Point. A four-wheel-drive vehicle, and food and water for at least two days are required to make this arduous road trip.

The sharply upturned beds of Upheaval Dome may be seen up the canyon from Upheaval Bottom at about Mile 44. It is a salt dome. Some would say that salt flowage was triggered by meteorite impact some 65 million years ago. Others would say "Fat chance!"

WHITE RIM SANDSTONE

An unusually straight stretch of river extends for nearly three miles below Fort Bottom at Mile 39.2. There is a nice display of the chocolate brown, lower Moenkopi Formation along the right bank where numerous gullies provide three-dimensional views of the bedding. There is a strange angular unconformity within the brown beds, a feature found elsewhere only at the edge of the White Rim sand bar in Elaterite Basin. Just downstream the White Rim Sandstone rises abruptly to the surface. The brown mudstones have draped across a north-trending, bar-shaped hump at the top of the White Rim, have been eroded to a plain, and then buried by more flat-bedded, brown mud. The straight course of the river parallels the flank of the hump, at least as far as the next bend in the river.

When the White Rim Sandstone rises to the surface, it is apparent that the large-scale cross bedding was deposited as windblown dunes. As the river cuts downward through the formation, however, a prominent horizontal bedding surface appears. Below that, the nature of the cross bedding changes to smaller, more wispy bedding that appears to be waterlaid. Then the smaller cross beds work their way upward, higher into the formation. We have apparently crossed a shoreline preserved in the White Rim Sandstone.

West wall of lower Labyrinth Canyon at Mile 37. The upturned beds in the lower cliff have been eroded and overlain by higher horizontal beds of the Moenkopi, forming an angular unconformity. High cliffs on the left consist of the Wingate Sandstone, capped by the Kayenta Formation.

Perhaps this is even a "fossil" barrier bar, such as Padre Island or Galveston Island on the Texas Gulf coast. By Anderson Bottom, the lower beds of the White Rim Sandstone have changed to the crinkly flat beds of sandy tidal flat environments as at the foot of Shafer Trail. Regional significance of this feature is discussed in Chapter Seven.

STILLWATER CANYON

By Powell's definition, Stillwater Canyon begins where the White Rim Sandstone rises to the surface at Mile 36. The appearance of the canyon begins to change, as the White Rim Sandstone gradually takes over the role of rimrock from the Wingate-Kayenta cliffs up stream. And, although Powell didn't realize it, the rocks of Labyrinth Canyon are of Triassic and Jurassic age (Mesozoic Era), and the rocks of Stillwater Canyon are of Permian age (Paleozoic Era) (see rock chart inside either cover). So there are some changes in the character of the canyons after all.

The semi-circular valley back of Anderson Bottom is an abandoned meander of the river. At some time in the not too distant past, perhaps a million years ago, the river flowed into the upstream approach and around the central butte, and then reentered its present course. Erosion along the river banks at the outside of the two sharp turns finally broke through the "gooseneck," shortening the river and leaving the meander hopelessly stranded. These entrenched abandoned meanders are sometimes called rincons (a Spanish word for "corner" or "nook").

There are a number of cliff dwellings at the base of the White Rim Sandstone below Anderson Bottom. The underlying red Organ Rock Shale weathers back readily, leaving protected coves at the contact between the formations. The change from porous sandstone above to impermeable shale below also localizes springs along the contact.

Downstream from Tuxedo Bottom (Mile 24) the red beds become more massive and cliff-forming, as the arkosic sandstones of the lower Cutler are exposed. White sandstone beds become more numerous in the section downstream, and discrete sandstone-filled channels are common. These are thin tongues and tidal channels filled with the white Cedar Mesa Sandstone near a zone of complex facies changes between the red arkoses of the Cutler from the east and white sands of the Cedar Mesa from the northwest. This zone of complex interfingering of different rock types trends across Canyonlands into the Needles District to the southeast.

A thin limestone that contains marine fossils rises to river level at Mile 16.7, representing the top of the Elephant Canyon Formation of Lower Permian age. The canyon walls from here to the confluence are composed of this formation. The marine limestones in the upper Elephant Canyon Formation are here interfingering with arkosic stream-deposited red beds from the east and the white coastal dunes of the Cedar Mesa Sandstone. This is a complex three-way facies change that centers around the confluence of the Green and Colorado Rivers (see Chapter Seven). As we approach the confluence, the river cuts ever deeper into the formation until most of the canyon walls are composed of limestone.

The Green River canyon turns upstream to the Colorado River, and

119

Lower Stillwater Canyon on the Green River. The lower cliffs and slopes are formed on marine sedimentary rocks of the Elephant Canyon Formation. The high cliffs are in the Cedar Mesa Sandstone. The confluence of the Green and Colorado rivers is "just around the bend."

the rocks fold abruptly upward at the confluence. The unconformity separating the Elephant Canyon Formation from underlying Pennsylvanian strata finally appears in the flexed rocks, and may be seen across the Colorado River, low in the canyon wall. Rocks of Pennsylvanian age have finally appeared at the surface, and they become plentiful downstream in Cataract Canyon. The flexure in the rocks marks the flank of Meander anticline, down which the Colorado River flows for several miles.

Most rivers merge such that both tributaries flow toward each other at a low angle; in other words confluences are generally Y-shaped in map view. However, the confluence of the Green and Colorado rivers is a maverick, as the course of the Green River heads up the course of the Colorado River at the juncture. Such a hook-shaped junction makes the Green River a "barbed tributary." There is no clue why this has happened. Barbed tributaries usually result from one stream (Green) having been captured by the other (Colorado) in an act called "stream piracy." Whatever the cause, the course of the Green River has been disrupted at some time in the dim, dark past.

East canyon wall at the confluence of the Green and Colorado rivers. There are two angular unconformities visible at the arrows, both are believed to have formed as the result of up-tilting of the layered rocks by salt flowage along Meander anticline.

121

Dead Horse Point seen from the Colorado River at Shafer Wash. Inner canyon cliffs just above the river are in the Elephant Canyon Formation, and the middle ledgy cliffs are in the upper Cutler Group. The high slopes include the Moenkopi and Chinle Formations. High cliffs containing the large cove are the Wingate Sandstone, capped by a thin remnant of Kayenta Formation. The Dead Horse Point shelter is visible atop the far right corner of the point.

CHAPTER 16
MEANDER CANYON

The Colorado River flows generally southwestward from Moab to the confluence, a distance of 64 miles, and like the Green River, it is sluggish above Cataract Canyon, dropping at a rate of barely a foot per mile. The canyon of the Colorado River above the confluence has never been formally named, probably because Major Powell never saw it, and who else would dare name a canyon? It seems appropriate to call it "Meander Canyon" for one of its best attributes.

The river first crosses the Moab salt structure as if it did not exist, and leaves through the upturned Chinle and Wingate formations at The Portal. From there it wanders around through King's Bottom syncline in the Glen Canyon Group before tackling the Cane Creek anticline, a salt-pillowed structure that trends northwestward, paralleling other salt anticlines in the area. There was not enough salt present to pierce the overlying rocks as at Moab, but salt flowage bulged the rocks upward, to form the Cane Creek anticline. Contorted bedding due to salt flowage hampered underground mining operations in the mine at Potash, so salt has been solution-mined since 1971. The Cutler red beds thin by about half across the structure, indicating that much of its growth was during Permian time.

The river challenges Cane Creek anticline, and cuts a canyon almost directly at right angles to the fold. Limestones of the Elephant Canyon Formation (Lower Permian Period), some containing large fossil clams, are exposed by erosion over the structure, and comprise the canyon walls downstream from the boat launching ramp below Potash. The topmost bed of the Honaker Trail Formation of Late Pennsylvanian age is exposed at river level on the crest of the anticline. Here, too, a well was drilled for oil during 1925-27, reaching a depth of 5,000 feet in salt. The drilling rig was brought down the river from Moab by barge, and was supplied by wagons over the often frozen river. Remains of the cable-tool drilling rig may be seen on the right bank.

Dead Horse Point, perched high atop the magnificent Wingate-Kayenta cliffs, may be seen as the river heads directly into Shafer Dome. This is another salt-bulged anticline, but the fold has a peculiar orientation. It parallels the northeasterly Cataract basement lineament to the east, but turns to parallel the Cane Creek and Moab structures to the west. The curved anticline seems to overlie a "trap door" faulted structure in the basement. In the meantime, the river first crosses the flank of the anticline, then flows along the axis, and finally takes a left, leaving Shafer Dome as awkwardly as it entered.

Uppermost limestones of the Elephant Canyon Formation come back down nearly to river level after the canyon leaves Shafer Dome, but then rise again as the river crosses the northwesterly plunge of the Lockhart anticline at about Mile 27. The limestones dip beneath the surface under the Grays Pasture syncline (Mile 23), only to reappear soon. Thin tongues of white Cedar Mesa Sandstone may be seen within red beds of the lower Cutler, as the river approaches the Cutler-Cedar Mesa facies change near the confluence.

A snow-white bed of gypsiferous sandstone punctuates the red, layered rocks in the vicinity of Sheep Bottom (Mile 19). Most folks will call this gypsum, but it is really sandstone. Then just around the corner, the strata dip noticeably toward both banks of the river, as we enter Meander anticline and another controversy.

MEANDER ANTICLINE

Early geologists mapping the area noticed that an anticlinal structure, a sharp upfold, generally follows the course of the Colorado River. Because of the apparent close relationship of the river's course and the structure, the fold was named "Meander anticline." They postulated that canyon-cutting had lightened the overburden, allowing the salt at shallow depth to rise into the resulting trend of lower pressure. This is certainly possible, but it is probably a much too simplistic explanation. Some folks explain the origin of all salt-intruded anticlines by this process, which is in most cases absurd. So what's wrong with the notion?

In the first place, the Colorado River system crosses far more salt structures than it follows! Most of the major salt valleys are dissected by one or another of the rivers without discretion, a pattern that surely does not fit the concept of erosional unloading.

Second, the salt structures were growing from Middle Pennsylvanian through Jurassic time. There are two notable angular unconformities in the vicinity of the confluence, apparently formed by salt flowage in Meander anticline late in Pennsylvanian, and during Lower Permian time respectively. If that is correct, salt was flowing in Meander anticline some 250 million years before the river could have possibly taken its present course. Salt flowage had ceased prior to Cretaceous time, when an inland sea covered the region with thousands of feet of Mancos Shale and the Mesaverde Group. The canyons could not have been established until the seas withdrew, and that was long after the salt stopped flowing.

Could there be any other objections to the unloading theory? The course of the canyon and Meander anticline are not closely coincident, especially at The Loop. There the course of the canyon definitely meanders around the fold, rather than the fold following the river.

Now for the biggest question of all: **Why is there no salt anticline under the Green River upstream from the confluence?** If unloading were the only process involved, Meander anticline should split and follow both rivers.

There is more to the origin of Meander anticline than erosional unloading of the salt along the canyon. The river has cut its canyon along the structure, not the other way around. So, like the proverbial egg, the structure came first, and the river, like the chicken, came later!

Now that we have seemingly beaten the Meander anticline controversy to a pulp, we can proceed into Cataract Canyon with some dignity.

North wall of Meander Canyon on the Colorado River at the mouth of Elephant Canyon three miles above the confluence with the Green River. The rocks, except for the highest cliffs of Cedar Mesa Sandstone, are in the Elephant Canyon Formation, named for this tributary. The lovely sand beach has long-since washed away to Lake Powell, but it will undoubtedly someday reappear.

CHAPTER 17
CATARACT CANYON

Below the confluence of the Green and Colorado rivers, the flow of the Colorado roughly doubles in size. The river gurgles and boils in a peculiar anticipation of what lies ahead in Cataract Canyon. This is the deepest and most awesome expression of Canyonlands Country; in this stretch the river can be vicious and relentless!

GRADIENT

Four miles below the confluence, the river changes dramatically. The gradient increases from about one foot per mile to an average of 10 feet per mile; it reaches a maximum of 38 feet per mile just above Lake Powell. And the rapids are almost continuous in some stretches. But why here?

As with most things in geology, the reasons for the sudden change in the river's attitude are complex and varied. Several explanations have been offered, most are partially or totally in error. The most plausible explanation involves deep-seated basement structure. It is at the first rapids that an ancient fault crosses the canyon deep underground, extending from the Abajo Mountains, through Chesler Park, Spanish (Cataract) Bottom, and on to the northwest to the San Rafael Swell. The increased gradient (nick point) is coincidental with the basement fault, and may be controlled by the fault.

Whatever the cause, the steepened gradient greatly increases the flow rate of the river. In recent years of deep snowpack and high runoff, flows have been measured at nearly 120,000 cubic feet per second (cfs) in Cataract Canyon. With such high water, 45-foot-long river rafts were flipped end-over-end, and 50-foot-tall cottonwood trees came shooting straight up from the eddies. Logjams nearly blocked the headwaters of Lake Powell downstream. It is likely that spring floods were much worse prior to the construction of dams upstream. Flow rates may have been double those of today. Prior to dam construction, a flow of more than 300,000 cfs was measured in Grand Canyon. It is unimaginable what Cataract Canyon rapids would be like during such floods.

RAPIDS

Rapids form where there are constrictions to the normal flow of water. Most rapids occur where mudflows enter the channel from short, steep tributary canyons after severe rain storms. Boulders of any size are carried

Lower Big Drop Rapid, sometimes called Kolb Rapid, in Cataract Canyon. Boulders caved from the crumbled canyon walls obstruct the flow of the Colorado River, creating a river-runners nightmare.

Threading "Satan's Gut" in Lower Big Drop Rapid, Cataract Canyon. Here we learn to HOLD ON! The boat, a so-called "rubber baloney," is 33 feet long for scale.

with the mud avalanche, partially damming the main channel. But this usual situation is not often the case in Cataract Canyon.

Rapids have formed in Cataract Canyon where the main channel has been severely restricted by mass movement of the canyon walls and by rock falls from the unstable cliffs. Massive blocks of bedrock are creeping toward the canyon on the slippery upper layers of rock salt buried at shallow depths. As the huge fractured blocks creep slowly but relentlessly from The Grabens area in the Needles District of Canyonlands National Park toward the canyon, entire sections of the canyon walls slump and rotate into the void of the canyon. They crowd the river's path into the opposite bank, severely restricting normal flow. Large blocks of rock promiscuously fall into the river from the broken up, slumping canyon walls, further complicating the issue. These obstructions cause large waves and holes in the river that may be hazardous to life, limb, and rubber boats.

Thus, nearly continuous rapids have formed throughout the canyon. In high-water stages, the waves and holes tend to get bigger, and the closely spaced rapids of Cataract Canyon tend to merge into long ones. Mile-long Rapid and the three Big Drop rapids are good examples. The ride may not be so thrilling in low water stages, and the rapids become rock gardens that are difficult to negotiate.

GYPSUM PLUGS

There is a series of well-defined gypsum plugs in Cataract Canyon, all lined up in a row along the Meander anticline. They have flowed upward from salt beds of the Paradox Formation, here at shallow depth, and undoubtedly contained salt when they were emplaced. None has pierced very high into the overlying strata, because little salt was deposited here. There is only a thousand feet of bedded salt beneath the Confluence, in contrast to about 5,000 feet at Potash near Moab, and it thins rapidly toward the southwest to nothing at Gypsum Canyon.

The first of the gypsum plugs in Cataract Canyon, Prommel Dome, is seen at the mouth of Lower Red Lake Canyon, and probably underlies the broad, circular floor of Spanish (Cataract) Bottom as well. As we just saw, it occurs at the intersection of the Cataract lineament and a northwest-trending fault, and was probably localized by the deep-seated basement structures.

The second gypsum plug, Harrison Dome, underlies Tilted Park at the mouths of Y Canyon and Cross Canyon. It is not nearly as well exposed as Prommel Dome, but is readily seen to have bowed up the overlying rocks to give Tilted Park its name. There is no apparent reason for its exact location, except that it rises above the Meander anticline salt structure.

Another gypsum plug, mapped as Crum Domes, occurs at the mouth of the next unnamed canyon down the river. It (or they, depending on how one maps the feature) is very well exposed like Prommel Dome, and like Harrison Dome there is no obvious reason for its exact location. A stroll through either Crum or Prommel Domes, via the canyons that dissect both structures, gives one a good look at the highly contorted nature of the flowed gypsum beds.

Lower Red Lake Canyon at Mile 213 on the Colorado River, just above the first rapids in Cataract Canyon. The jumbled mass of rock at the mouth of the tributary canyon contains gypsum, squeezed upward from the Paradox Formation at shallow depth along the larger Meander anticline.

A quiet stretch of the Colorado River in the heart of Cataract Canyon. The layered rocks forming the lower two-thirds of the canyon walls are in the Honaker Trail Formation. The higher dark-colored cliffs and slopes are in the Elephant Canyon Formation, and the high cliffs are Cedar Mesa Sandstone. Note the slump block of collapsed canyon wall on the lower right of the photo.

Exposures of the Paradox Formation near river level, from the Big Drop Rapids downriver to Gypsum Canyon, are undisturbed beds of gypsum. Thin layers of shattered rocks (collapse breccias) occur within the gypsum beds, indicating that some salt had been present and dissolved, causing overlying beds to collapse and break up during recent exposure. These are the deepest and oldest rocks exposed in Cataract Canyon.

THE CLIFFS

Above river level, Cataract Canyon consists of a series of drab gray cliffs of limestones, with interbedded, equally as drab, marine sandstones and shales. Most of these canyon walls represent the Honaker Trail Formation, the upper formation of the Hermosa Group of Middle to Late Pennsylvanian age. The obviously layered rocks were deposited in the 300-million-year-old, more or less, seaway that lay across what is now Canyonlands Country (see Chapter Six). The rocks are rich in fossils of critters that lived in the sea during those days; crinoids, brachiopods, bryozoa, and microfossils dominate the fossil assemblage, but corals, clams, snails and even trilobites may be found. Almost every bed contains fossils if one looks hard enough, and they are the tools for dating and correlating these strata. The sea withdrew from the region, and erosion again beveled the surface near the end of Pennsylvanian time, forming a widespread unconformity. The erosional surface may represent 10 million years of Earth history.

When the sea returned in Early Permian time, it came from another direction. This time an embayment of the Oquirrh Sea of north-central Utah spilled into what would be Canyonlands Country, and the Elephant Canyon Formation was deposited. The best exposures of these strata occur near the confluence. Elephant Canyon, a tributary of the Colorado River three miles above the confluence, is the reference section for the formation. East of the confluence, the marine rocks thin and interfinger with stream and dune deposits of the Cutler Formation. South of the confluence, the marine Elephant Canyon Formation thins and interfingers with red stream and tidal flat beds of the Halgaito Shale, at the base of the Cutler Group (see Chapter Seven). And high above the Confluence, limestones of the Elephant Canyon Formation are interfingering upward and laterally with coastal sands of Cedar Mesa Sandstone.

Thus, the area around the confluence is a "hornet's nest" of stratigraphic complexities that no one really understands in detail. The resulting rocks form the higher canyon walls, seen throughout Cataract Canyon from the confluence downriver to Hite on upper Lake Powell. The rimrock along the entire length of the canyon is the Cedar Mesa Sandstone of Lower Permian age. Younger formations have been stripped well back from Cataract Canyon by erosion, and seem to become nonexistent from this viewpoint.

With a closeup look at the rocks and times of the canyons of the Green and Colorado rivers, culminating in Cataract Canyon, we have plumbed the very heart and soul of Canyonlands Country. Now only the relentless downcutting of the river and endless time can heighten the joys found here.

SUGGESTED READING

Nontechnical:

The Colorado Plateau: A Geologic History, 1983, Baars, Donald L.: University of New Mexico Press, Albuquerque, NM, 279 pages.

A River Runners' Guide to Cataract Canyon and Approaches, 1987, Baars, Don: Cañon Publishers Ltd., Evergreen, CO, 80 pages.

Geology of Canyonlands and Cataract Canyon [technical in part] 1971, Baars, D.L. and Molenaar, C.M.: Four Corners Geological Society, Durango, CO, 99 pages.

Geologic History of Utah, 1988, Hintze, Lehi F.: Brigham Young University Geology Studies Special Publication 7, 202 pages.

The Geologic Story of Canyonlands National Park, 1974, Lohman, Stanley W.: U.S. Geological Survey Bulletin 1327, 126 pages.

The Geologic Story of Arches National Park, 1975, Lohman, Stanley W.: U.S. Geological Survey Bulletin 1393, 113 pages.

Geology of Utah, 1986, William Lee Stokes: Utah Museum of Natural History and Utah Geological and Mineral Survey Occasional Paper 6, 280 pages.

Geological Guidebooks [mostly technical]:

Cataract Canyon and Vicinity, 1987, John A. Campbell, editor: Four Corners Geological Society 10th Field Conference Guidebook, 200 pages.

Geology of the Paradox Basin, 1981, Del L. Wiegand, editor: Rocky Mountain Association of Geologists Field Conference Guidebook, 285 pages.

Permianland, 1979, D.L. Baars, editor: Four Corners Geological Society 9th Field Conference Guidebook, 186 pages.

Canyonlands Country, 1975, James E. Fassett, editor: Four Corners Geological Society 8th Field Conference Guidebook, 281 pages.

Technical:

The technical geological literature is voluminous. For the serious reader, the bibliography in *The Colorado Plateau: A Geologic History* (Baars, 1983) is relatively complete and current.

GLOSSARY
(Modified from Stokes, 1966)

angular unconformity, n. An unconformity or break between two series of rock layers such that rocks of the lower series underlie rocks of the upper series at an angle; the two series are not parallel. The lower series was deposited, then tilted and eroded prior to deposition of the upper layers.

anticline, n. An elongate fold in the rocks in which sides slope downward and away from the crest; an upfold.

arkose, n. A sandstone containing a significant proportion of feldspar grains, usually signifying a source area composed of granite or gneiss. Rocks of the Cutler Formation, especially east of Moab, are very "arkosic," containing a high proportion of pink feldspar sand grains.

basement, n. In geology, the crust of the Earth beneath sedimentary deposits, usually, but not necessarily, consisting of metamorphic and/or igneous rocks of Precambrian age.

basement fault, n. A fault that displaces basement rocks and originated prior to deposition of overlying sedimentary rocks. Such faults may or may not extend upward into overlying strata, depending upon their history of rejuvenation.

base level, n. The level, actual or potential, toward which erosion constantly works to lower the land. Sea level is the general base level, but there may be local, temporary base levels such as lakes.

bentonite, n. A rock composed of clay minerals and derived from the alteration of volcanic tuff or ash.

brachiopod, n. A type of shelled marine invertebrate now relatively rare but abundant in earlier periods of Earth history. They are common fossils in rocks of Paleozoic age. Brachiopods have a bivalve shell that is symmetrical right and left of center.

bryozoa, n. Tiny aquatic animals that build large colonial structures that are common as fossils in rocks of Paleozoic age.

chert, n. A very dense rock composed of fine-grained silica (quartz - SiO_2) usually found as irregularly-shaped nodules or as distinct beds in limestones. Very hard, it breaks in conchoidal (curved) sheets much like glass. Depending on the color, which results from contained impurities, it is variously known as flint or Jasper, and other odd names, especially by rockhounds. Jasper is red chert containing iron-oxide impurities, often replacing fossils in Canyonlands Country.

clastic rocks, n. Deposits consisting of fragments of preexisting rocks; conglomerate, sandstone, and shale are examples.

conglomerate, n. The consolidated equivalent of gravel. The constituent rock and mineral fragments may be of varied composition and range widely in size. The rock fragments are rounded and smoothed from transportation by water.

contact, n. The surface, often irregular, which constitutes the junction of two bodies of rock.

continental deposits, n. Deposits laid down on land or in bodies of water not connected with the ocean.

correlation, n. The process of determining the position or time of occurrence of one geologic phenomenon in relation to others. Usually it means determining the equivalence of geologic formations in separated areas through a comparison and study of fossils or rock peculiarities.

crinoid, n. Marine invertebrate animals, abundant as fossils in rocks of Paleozoic age. Most lived attached to the bottom by a jointed stalk, the "head" resembling a lilylike plant, hence the common name "sea lily."

dike, n. A sheetlike body of igneous rock that filled a fissure in older rock while in a molten state. Dikes that intrude layered rocks cut the beds at an angle.

disconformity, n. A break in the orderly sequence of stratified rocks above and below which the beds are parallel. The break is usually indicated by erosional channels, indicating a lapse of time or absence of part of the rock sequence.

dolomite, n. A mineral composed of calcium and magnesium carbonate, or a rock composed chiefly of the mineral dolomite, formed by alteration of limestone.

dome, n. An upfold in which strata dip downward in all directions from a central area; the opposite of a basin.

eolian, adj. Pertaining to wind. Designates rocks or soils whose constituents have been transported and deposited by wind. Windblown sand and dust (loess) deposits are termed "eolian."

erosional unconformity, n. A break in the continuity of deposition of a series of rocks caused by an episode of erosion.

extrusive rock, n. A rock that has solidified from molten material poured or thrown out onto the Earth's surface by volcanic activity.

facies, n. Generally, the term "facies" refers to a physical aspect or characteristic of a sedimentary rock, as related to adjacent strata. It is usually applied to distinguish different aspects of the sediments in time equivalent or laterally continuous beds. For example, the white sandstone "facies" of the Cedar Mesa Sandstone changes laterally to the age-equivalent red sandstone "facies" of the Cutler Group in Canyonlands Country. Such a change from one aspect to another is called a "facies change."

feldspar, n. Complex aluminum silicate minerals that usually occur as rectangular white or pink crystals in granite, gneiss, and other rocks formed under high temperatures. Commonly occurs as sand grains, along with quartz and other sand-size particles, when eroded from the parent rock, and if present in considerable quantity the rock is then called "arkose" or "arkosic sandstone." Rocks of the Cutler Formation, especially east of Moab, are very "arkosic," containing high proportions of pink feldspar grains.

fault, n. A break or fracture in rocks, along which there has been movement, one side relative to the other. Displacement along a fault may be vertical (normal or reverse fault) or lateral (strike-slip or "wrench" fault).

Foraminifera, n. Generally microscopic one-celled animals (Protozoa), almost entirely of marine origin, with sufficiently durable shells capable of being preserved as fossils. They are usually abundant in marine sediments, and are sufficiently small to be retrievable in drill cuttings and cores.

formation, n. The fundamental unit in the local classification of layered rocks, consisting of a bed or beds of similar or closely related rock types, and differing from strata above and below. A formation must be readily distinguishable, thick enough to be mappable, and of broad regional extent. A formation may be subdivided into two or more members, and/or combined with other closely related formations to form a group.

fusulinid, n. Extinct, marine, one-celled animals (Foraminifera) that are usually abundant in marine sedimentary rocks of Pennsylvanian and Permian age. Because the species evolved very rapidly with geologic time, they are very useful in dating rocks of late Paleozoic age. They consist of enrolled, elongate chambers that form spindle-shaped, microscopic fossils.

geologic map, n. A map showing the geographic distribution of geologic formations and other geologic features, such as folds, faults, and mineral deposits, by means of color or other appropriate symbols.

gneiss, n. A banded metamorphic rock with alternating layers of usually tabular, unlike minerals.

granite, n. An intrusive igneous rock with visibly granular, interlocking, crystalline quartz, feldspar, and perhaps other minerals.

halite, n. Common rock salt (NaCl); precipitates from sea water under conditions of intense evaporation. Occurs as bedded, usually crystalline, white or red rock only in the subsurface; flows plastically under high confining pressure.

igneous rock, n. Rocks formed by solidification of molten material (magma), including rocks crystallized from cooling magma at depth (intrusive), and those poured out onto the surface as lavas (extrusive).

intrusive rock, n. Rock that has solidified from molten material within the Earth's crust and did not reach the surface; usually has a visibly crystalline texture.

law of superposition, n. A concept stating that, if undisturbed, any sequence of sedimentary rocks will have the oldest beds at the base and the youngest at the top.

limestone, n. A bedded sedimentary deposit consisting chiefly of calcium carbonate, usually formed from the calcified hard parts of organisms.

metamorphic rock, n. Rocks formed by the alteration of preexisting igneous or sedimentary rocks, usually by intense heat and/or pressure, or mineralizing fluids.

microfossil, n. Any fossil too small to be studied without magnification.

orogeny, n. Literally, the process of formation of mountains, but practically it refers to the processes by which structures in mountainous regions were formed, including folding, thrusting, and faulting in the outer layers of the crust, and plastic folding, metamorphism and plutonism (emplacement of magmas) in the inner layers. An episode of structural deformation may be called an orogeny, e.g., the Laramide Orogeny.

Reef, n. Some topographic features are called "reefs" in Canyonlands Country, such as Capitol Reef, and San Rafael Reef, named by pioneers in the 1800s. They thought of their horse-drawn wagons as "prairie schooners" and severe obstacles to travel as "reefs" synonymously with nautical terms. These "reefs" are jagged ridges of sharply upturned Navajo and Wingate sandstone along monoclinal folds.

ripple marks, n. A series of small ridges and troughs formed on a sandy surface by wind or water currents, or by wave action. They are often preserved on bedding planes in sedimentary rocks.

salt dome, n. Usually a circular uplift of sedimentary rocks caused by the pushing upward of a mass of salt and/or gypsum at depth. As in eastern Canyonlands Country, the circular cells may coalesce into elongate walls aligned along deep-seated faults to form salt anticlines.

sandstone, n. A consolidated rock composed of sand grains cemented together; usually composed predominantly of quartz, it may contain other sand-size fragments of rocks and/or minerals.

schist, n. A platy crystalline metamorphic rock that splits into thin flakes or slabs. Minerals are elongated and flattened from intense compression of the original rock.

sedimentary rock, n. Rocks composed of sediments, usually aggregated through processes of water, wind, glacial ice, or organisms, derived from preexisting rocks, or in the case of limestones, constituent particles are usually derived from organic processes.

shale, n. Solidified muds, clays, and silts, that are fissile (split like paper) and break along original bedding planes.

sill, n. A tabular body of igneous rock that was injected in the molten state concordantly between layers of preexisting rocks.

stratigraphy, n. The definition and interpretation of the layered rocks, the conditions of their formation, their character, arrangements, sequence, age, distribution, and correlation, using fossils and other means.

stratum, n. A single layer of sedimentary rock, separated from adjacent strata by surfaces of erosion, non-deposition, or abrupt changes in character. Plural strata.

syncline, n. An elongate, troughlike downfold in which the sides dip downward and inward toward the axis.

tectonic, adj. Pertaining to rock structures formed by Earth movements, especially those that are widespread.

trilobite, n. A general term for a group of extinct animals (arthropods) that occurs as fossils in rocks of Paleozoic age. They consist of flattened, segmented shells with a distinct thoraxial lobe and paired appendages; usually found as partial fragments.

type locality, n. The place from which the name of a geologic formation is taken, and where the unique characteristics of the formation may be examined.

unconformity, n. A surface of erosion or nondeposition separating sequences or layered rocks.

upwarp, n. A broad area where the layered rocks have been uplifted by internal forces, a classic example is the Monument Upward in Canyonlands Country.

INDEX